Contemporary Discourse in the Field of BIOLOGY™

The Growth and Development of Specialized Cells, Tissues, and Organs

An Anthology of Current Thought

Edited by Craig C. Freudenrich, Ph.D.

The Rosen Publishing Group, Inc., New York

This book is dedicated to my wife, Theresa, for her understanding and patience during this project

Published in 2006 by The Rosen Publishing Group, Inc.
29 East 21st Street, New York, NY 10010

First Edition

Library of Congress Cataloging-in-Publication Data

The growth and development of specialized cells, tissues, and organs: an anthology of current thought / edited by Craig C. Freudenrich.
p. cm. — (Contemporary discourse in the field of biology)
Includes bibliographical references and index.
ISBN 1-4042-0401-6 (library binding)
1. Cell differentiation.
I. Freudenrich, Craig C. II. Series.
QH607.G76 2006
571.8'35—dc22

2004026157

Manufactured in the United States of America

On the cover: Bottom right: A microscopic observation of a mouse forelimb during embryonic development, in which the expression of a subfamily of proteins known as Glis is analyzed for its role in regulation of developmental processes. Top: Digital cell. Far left: Digital cell. Bottom left: Austrian monk and botanist Gregor Johann Mendel (1822–1884).

CONTENTS

Introduction

Life on Earth originated approximately 3,700 to 3,800 million years ago. Early life-forms were simple, single-celled organisms (bacteria) that thrived alone and in communities for more than 3,000 million years. About 700 million years ago, the first multicelled organisms appeared. Multicelled organisms were the pretext for the rapid evolution of complex life-forms that occurred between 500 and 600 million years ago during the Cambrian period of geologic time. From this period to the present, life-forms diversified into many millions of species of animals and plants.

Regardless of their complexity, all living things must carry out the same biological functions, such as metabolism, growth, reproduction, and response to stimuli. As living things evolved into multicellular forms and grew larger and more complex, various cells became specialized, in that they became adapted to carry out specific functions. For example, nerve cells in animals transmit electrochemical signals; nematocysts, or stinging cells, in invertebrates are specialized for defense;

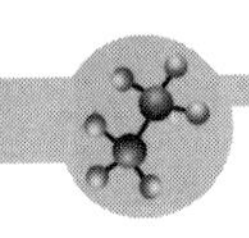

and guard cells in plants regulate the openings of pores to control the exchange of gases in the leaves for photosynthesis. Furthermore, specialized cells of similar functions group together to form tissues (nerves, muscles, xylem, and phloem). Various tissues combine into organs (heart, kidney, brain, root, and leaf), and various organs work together to form organ systems (nervous system, cardiovascular system, digestive system, and reproductive system).

All organisms begin as single cells, whether the product of asexual reproduction (mitosis) or sexual reproduction (union of egg and sperm cells in fertilization to produce a zygote). From these simple origins, complex life such as trees, fish, and humans form. Let's look at humans as an example. Over the course of development, the fertilized egg, or zygote, will transform from its single-celled beginning through several stages into a person containing trillions of cells. The stages are as follows:

- Cleavage—After fertilization, the zygote rapidly divides many times to form a hollow ball of cells called a blastula. The cells located at the animal pole of the blastula are larger than the cells at the other end, or vegetal pole. Besides creating a multicelled embryo, cleavage organizes the embryo for later development.
- Gastrulation—Cells in various parts of the blastula divide at different rates. Various

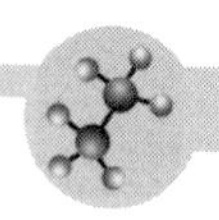

cells also begin to move within the gastrula, ultimately forming an embryo with three distinct layers of cells (the endoderm, mesoderm, and ectoderm) and the fluid-filled cavity known as the archenteron.

- Organ formation/pattern formation—From the various layers of cells, organs are formed. The endoderm gives rise to digestive, renal, and reproductive organs. The mesoderm gives rise to skeletal, muscle, and circulatory organs. The ectoderm gives rise to skin, eyes, and nervous system organs. A process called induction, in which a group of cells elicits changes in a neighboring group of cells, is also important in forming organs. Furthermore, during this period, patterns such as head end and tail end, front and back, and right and left are formed as well as various limbs, such as arms and legs.
- Growth and further organ formation—The embryo continues to form organs, grow, and strengthen those organs throughout the remainder of gestation until birth.

The processes important in the formation and subsequent growth of the embryo include cell division, changes in cell shape, migrations of cells, specialization or differentiation, and programmed cell death, or apoptosis.

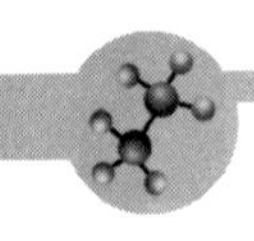

How does this transformation come about? How are these cellular processes—division, migration, differentiation, apoptosis—coordinated to produce a muscle cell in one part, a nerve cell in another part, and more complex forms and patterns such as the brain or a limb? Likewise, how does a seedling know at which end to develop leaves and at which end to develop roots? As all cells in an organism possess the same genetic information (the information to produce a whole organism), how is some of that information selected in one type of cell, yet silenced in another? Can all cells be used potentially to develop into an organism or a particular organ, or is that capability reserved for only a few? These are but a few of the questions that are central to the field of developmental biology. To examine these issues in developmental biology, scientists study a number of organisms such as the fruit fly (*Drosophila*), a soil worm (*Caenorhabditis elegans*), the frog (*Xenopus*), chick embryos, and plants belonging in genus *Arabidopsis*. These organisms have the advantages of small genomes, of which many have been fully sequenced, and short development durations. In addition to whole organisms, individual organs such as muscle, liver, and leaf hairs (trichomes) can serve as useful models for development. The articles in this anthology will address some of the major issues in developmental biology and use some of the model systems for development that we have mentioned so far.

In chapter 1, we examine what it means to be specialized by reviewing skeletal muscle cells as an example. How is muscle specialized? How can scientists

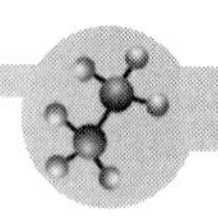

manipulate the genetic information within muscle to grow new cells? Furthermore, must all complex life-forms be made of many specialized cells? We see how a marine alga (*Caulerpa*) represents an apparent contradiction to this idea.

In chapter 2, we examine various biochemical and cellular processes in differentiation. We start by looking at plant trichomes and how this system can be used to model all aspects of differentiation in plants. How is differentiation in plants different than in animals? What are the relationships between cell shape, cell division, and differentiation in plants? We see how a type of ribonucleic acid (RNA) censors information in the genome to select which information gets expressed and which does not. How do differential gene regulation and biochemical gradients cause limbs to be formed and patterns—specifically right from left—to develop in animals? Finally, how does apoptosis contribute to development?

In chapter 3, we look at normal and abnormal modes of differentiation. How is it that the liver can regenerate, while other organs cannot? We see how lessons learned from developmental biology are encouraging scientists to take a new look at cancer. Is cancer an abnormal process of differentiation rather than a process of abnormal cell growth? Are biochemical gradients important in tumor formation? Are mechanical factors, which are important in tissue differentiation, disrupted in cancer, thereby leading to tumor formation?

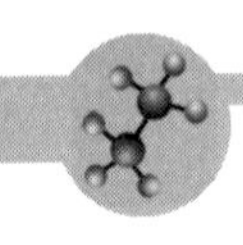

Finally, in chapter 4, we examine artificial means of cell differentiation. How can we use cell/molecular techniques to engineer new tissues to replace damaged or failed ones? As tissue engineering involves the use of undifferentiated cells or stem cells, particularly embryonic stem cells, it has sparked many ethical issues and much political debate. What are the issues and current thoughts in this area? What, if anything, should be done to control this type of research? As each cell in an organism contains the genetic information to make a whole organism, has it been possible to clone entire organisms from single cells? Could humans be cloned? What are the problems and progress in this area of research? Even if we master the concepts of developmental biology and can make new tissues, organs, and even organisms, these newly engineered forms will still function as their natural counterparts. However, will it be possible to create synthetic forms of life with different functions so that they will be capable of completing tasks that have no natural counterparts? The topics outlined above should summarize key concepts of growth and development of specialized cells, tissues, and organs and provide you with a basis for further study in this fascinating field. —*CCF*

Specialization of Cells

1

A muscle cell is an example of a cell that is highly specialized for contraction. Skeletal muscle cells are large, multinucleated cells with ordered groups of overlapping thin and thick protein filaments. Upon stimulation, the protein cross-bridges form between the thin and thick filaments. The cross-bridges pull the thin filaments, which slide past the thick filaments and cause the muscle to shorten (isotonic contraction) or generate great amounts of force (isometric contractions). Other types of muscle, such as cardiac and smooth, have the same basic mechanisms for contraction but vary in size and ordered structures. For example, smooth muscles do not have the highly ordered filament groups that skeletal muscles do; this arrangement allows them to stretch. In contrast, cardiac muscle is intermediate between the skeletal and smooth muscle.

One idea in biology is that as a cell becomes specialized, it loses its ability to reproduce. As we see in the article by H. Lee Sweeney, this idea comes into question. By using gene therapy and injecting growth factors into existing muscle, Dr. Sweeney shows us that muscle cells can indeed reproduce, grow, and strengthen. His gene therapy experiments may be useful for treating degenerative muscle diseases, such as Duchenne muscular dystrophy. However, in the future, this approach may also be abused by athletes determined to improve their muscle mass and athletic performance for competition. —CCF

"Gene Doping"
by H. Lee Sweeney
***Scientific American*, July 2004**

Gene therapy for restoring muscle lost to age or disease is poised to enter the clinic, but elite athletes are eyeing it to enhance performance. Can it be long before gene doping changes the nature of sport?

Athletes will be going to Athens next month to take part in a tradition begun in Greece more than 2,000 years ago. As the world's finest specimens of fitness test the extreme limits of human strength, speed and agility, some of them will probably also engage in a more recent, less inspiring Olympic tradition: using performance-enhancing drugs. Despite repeated scandals, doping has become irresistible to many athletes, if only to keep pace with competitors who are doing it. Where winning is paramount, athletes will seize any opportunity to gain an extra few split seconds of speed or a small boost in endurance.

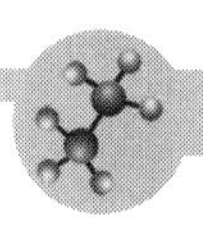

Sports authorities fear that a new form of doping will be undetectable and thus much less preventable. Treatments that regenerate muscle, increase its strength, and protect it from degradation will soon be entering human clinical trials for muscle-wasting disorders. Among these are therapies that give patients a synthetic gene, which can last for years, producing high amounts of naturally occurring muscle-building chemicals.

This kind of gene therapy could transform the lives of the elderly and people with muscular dystrophy. Unfortunately, it is also a dream come true for an athlete bent on doping. The chemicals are indistinguishable from their natural counterparts and are only generated locally in the muscle tissue. Nothing enters the bloodstream, so officials will have nothing to detect in a blood or urine test. The World Anti-Doping Agency (WADA) has already asked scientists to help find ways to prevent gene therapy from becoming the newest means of doping. But as these treatments enter clinical trials and, eventually, widespread use, preventing athletes from gaining access to them could become impossible.

Is gene therapy going to form the basis of high-tech cheating in athletics? It is certainly possible. Will there be a time when gene therapy becomes so commonplace for disease that manipulating genes to enhance performance will become universally accepted? Perhaps. Either way, the world may be about to watch one of its last Olympic Games without genetically enhanced athletes.

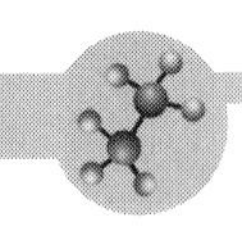

Loss Leads to Gain

Research toward genetically enhancing muscle size and strength did not start out to serve the elite athlete. My own work began with observing members of my family, many of whom lived well into their 80s and 90s. Although they enjoyed generally good health, their quality of life suffered because of the weakness associated with aging. Both muscle strength and mass can decrease by as much as a third between the ages of 30 and 80.

There are actually three types of muscle in the body: smooth muscle, lining internal cavities such as the digestive tract; cardiac muscle in the heart; and skeletal muscle, the type most of us think of when we think of muscle. Skeletal muscle constitutes the largest organ of the body, and it is this type—particularly the strongest so-called fast fibers—that declines with age. With this loss of strength, losing one's balance is more likely and catching oneself before falling becomes more difficult. Once a fall causes a hip fracture or other serious injury, mobility is gone completely.

Skeletal muscle loss occurs with age in all mammals and probably results from a cumulative failure to repair damage caused by normal use. Intriguingly, aging-related changes in skeletal muscle resemble the functional and physical changes seen in a suite of diseases collectively known as muscular dystrophy, albeit at a much slower rate.

In the most common and most severe version of MD—Duchenne muscular dystrophy—an inherited

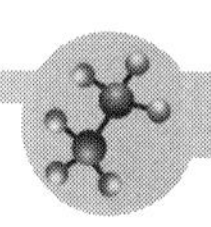

gene mutation results in the absence of a protein called dystrophin that protects muscle fibers from injury by the force they exert during regular movement. Muscles are good at repairing themselves, although their normal regenerative mechanisms cannot keep up with the excessive rate of damage in MD. In aging muscles the rate of damage may be normal, but the repair mechanisms become less responsive. As a result, in both aging and Duchenne MD, muscle fibers die and are replaced by infiltrating fibrous tissue and fat.

In contrast, the severe skeletal muscle loss experienced by astronauts in microgravity and by patients immobilized by disability appears to be caused by a total shutdown of muscles' repair and growth mechanism at the same time apoptosis, or programmed cell death, speeds up. This phenomenon, known as disuse atrophy, is still not fully understood but makes sense from an evolutionary perspective. Skeletal muscle is metabolically expensive to maintain, so keeping a tight relation between muscle size and its activity saves energy. Skeletal muscle is exquisitely tuned to changing functional demands. Just as it withers with disuse, it grows in size, or hypertrophies, in response to repeated exertions. The increased load triggers a number of signaling pathways that lead to the addition of new cellular components within individual muscle fibers, changes in fiber type and, in extreme conditions, addition of new muscle fibers.

To be able to influence muscle growth, scientists are piecing together the molecular details of how muscle is

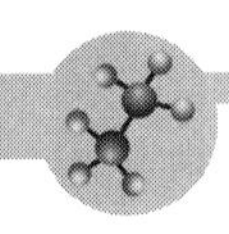

naturally built and lost. Unlike the typical cell whose membrane contains liquid cytoplasm and a single nucleus, muscle cells are actually long cylinders, with multiple nuclei, and cytoplasm consisting of still more long tiny fibers called myofibrils. These myofibrils, in turn, are made of stacks of contractile units called sarcomeres. Collectively, their shortening produces muscle contractions, but the force they generate can damage the muscle fiber unless it is channeled outward. Dystrophin, the protein missing in Duchenne muscular dystrophy patients, conducts this energy across the muscle cell's membrane, protecting the fiber.

Yet even with dystrophin's buffering, muscle fibers are still injured by normal use. In fact, that is believed to be one way that exercise builds muscle mass and strength. Microscopic tears in the fibers caused by the exertion set off a chemical alarm that triggers tissue regeneration, which in muscle does not mean production of new muscle fibers but rather repairing the outer membrane of existing fibers and plumping their interior with new myofibrils. Manufacturing this new protein requires activation of the relevant genes within the muscle cell's nuclei, and when the demand for myofibrils is great, additional nuclei are needed to bolster the muscle cell's manufacturing capacity.

Local satellite cells residing outside the muscle fibers answer this call. First these muscle-specific stem cells proliferate by normal cell division, then some of their progeny fuse with the muscle fiber, contributing their nuclei to the cell. Both progrowth and antigrowth factors are involved in regulating this process. Satellite cells

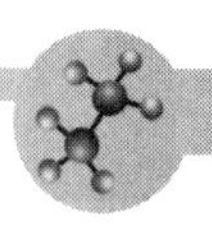

respond to insulinlike growth factor I, or IGF-I, by undergoing a greater number of cell divisions, whereas a different growth-regulating factor, myostatin, inhibits their proliferation.

With these mechanisms in mind, about seven years ago my group at the University of Pennsylvania, in collaboration with Nadia Rosenthal and her colleagues at Harvard University, began to assess the possibility of using IGF-I to alter muscle function. We knew that if we injected the IGF-I protein alone, it would dissipate within hours. But once a gene enters a cell, it should keep functioning for the life of that cell, and muscle fibers are very long-lived. A single dose of the IGF-I gene in elderly humans would probably last for the rest of their lives. So we turned our attention to finding a way to deliver the IGF-I gene directly to muscle tissue.

Donning New Genes

Then as now, a major obstacle to successful gene therapy was the difficulty of getting a chosen gene into the desired tissue. Like many other researchers, we selected a virus as our delivery vehicle, or vector, because viruses are skilled at smuggling genes into cells. They survive and propagate by tricking the cells of a host organism into bringing the virus inside, rather like a biological Trojan horse. Once within the nucleus of a host cell, the virus uses the cellular machinery to replicate its genes and produce proteins. Gene therapists capitalize on this ability by loading a synthetic gene into the virus and removing any genes the virus could use to cause disease or to replicate itself. We selected a tiny virus called adeno-associated virus (AAV) as our vector, in part because it infects human muscle readily but does not cause any known disease.

We modified it with a synthetic gene that would produce IGF-I only in skeletal muscle and began by trying it out in normal mice. After injecting this AAV-IGF-I combination into young mice, we saw that the muscles' overall size and the rate at which they grew were 15 to 30 percent greater than normal, even though the mice were sedentary. Further, when we injected the gene into the muscles of middle-aged mice and then allowed them to reach old age, their muscles did not get any weaker. To further evaluate this approach and its safety, Rosenthal created mice genetically engineered to overproduce IGF-I throughout their skeletal muscle. Encouragingly, they developed normally except for having skeletal muscles that ranged from 20 to 50 percent larger than those of regular mice. As these transgenic mice aged, their muscles retained a regenerative capacity typical of younger animals. Equally important, their IGF-I levels were elevated only in the muscles, not in the bloodstream, an important distinction because high circulating levels of IGF-I can cause cardiac problems and increase cancer risk. Subsequent experiments showed that IGF-I overproduction hastens muscle repair, even in mice with a severe form of muscular dystrophy.

Raising local IGF-I production allows us to achieve a central goal of gene therapy to combat muscle-wasting diseases: breaking the close connection between muscle use and its size. Simulating the results of muscle exercise in this manner also has obvious appeal to the elite athlete. Indeed, the rate of muscle growth in young

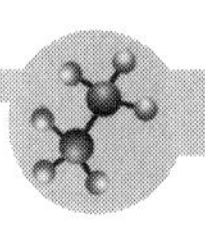

sedentary animals suggested that this treatment could also be used to genetically enhance performance of healthy muscle. Recently my laboratory worked with an exercise physiology group headed by Roger P. Farrar of the University of Texas at Austin to test this theory.

We injected AAV-IGF-I into the muscle in just one leg of each of our lab rats and then subjected the animals to an eight-week weight-training protocol. At the end of the training, the AAV-IGF-I-injected muscles had gained nearly twice as much strength as the uninjected legs in the same animals. After training stopped, the injected muscles lost strength much more slowly than the unenhanced muscle. Even in sedentary rats, AAV-IGF-I provided a 15 percent strength increase, similar to what we saw in the earlier mouse experiments.

We plan to continue our studies of IGF-I gene therapy in dogs because the golden retriever breed is susceptible to a particularly severe form of muscular dystrophy. We will also do parallel studies in healthy dogs to further test the effects and safety of inducing IGF-I overproduction. It is a potent growth and signaling factor, to which tumors also respond.

Safety concerns as well as unresolved questions about whether it is better to deliver AAV in humans through the bloodstream or by direct injection into muscle mean that approved gene therapy treatments using AAV-IGF-I could be as much as a decade away. In the shorter term, human trials of gene transfer to replace the dystrophin gene are already in planning stages, and the Muscular Dystrophy Association will soon begin a clinical trial of IGF-I injections to treat myotonic dystrophy, a condition that causes prolonged muscle contraction and, hence, damage.

A still more immediate approach to driving muscle hypertrophy may come from drugs designed to block myostatin. Precisely how myostatin inhibition builds muscle is still unclear, but myostatin seems to limit muscle growth throughout embryonic development and adult life. It acts as a brake on normal muscle growth and possibly as a promoter of atrophy when functional demands on muscle decrease. Experiments on genetically engineered mice indicate that the absence of this antigrowth factor results in considerably larger muscles because of both muscle fiber hypertrophy and hyperplasia, an excessive number of muscle fibers.

Making Muscle and More

Pharmaceutical and biotechnology companies are working on a variety of myostatin inhibitors. Initially, the possibility of producing meatier food animals piqued commercial interest. Nature has already provided examples of the effects of myostatin blockade in the Belgian Blue and Piedmontese cattle breeds, both of which have an inherited genetic mutation that produces a truncated, ineffective version of myostatin. These cattle are often called double-muscled, and their exaggerated musculature is all the more impressive because an absence of myostatin also interferes with fat deposition, giving the animals a lean, sculpted appearance.

The first myostatin-blocking drugs to have been developed are antibodies against myostatin, one of which may soon undergo clinical testing in muscular

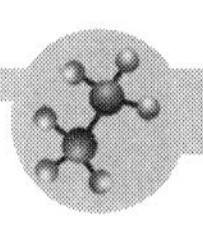

dystrophy patients. A different approach mimics the cattle mutation by creating a smaller version of myostatin, which lacks the normal molecule's signaling ability while retaining the structures that dock near satellite cells. This smaller protein, or peptide, essentially caps those docking locations, preventing myostatin from attaching to them. Injecting the peptide into mice produces skeletal muscle hypertrophy, and my colleagues and I will be attempting to create the same effect in our dog models by transferring a synthetic gene for the peptide.

Myostatin-blocking therapies also have obvious appeal to healthy people seeking rapid muscle growth. Although systemic drugs cannot target specific muscles, as gene transfer can, drugs have the benefit of easy delivery, and they can immediately be discontinued if a problem arises. On the other hand, such drugs would be relatively easy for sport regulatory agencies to detect with a blood test.

But what if athletes were to use a gene therapy approach similar to our AAV-IGF-I strategy? The product of the gene would be found just in the muscle, not in the blood or urine, and would be identical to its natural counterpart. Only a muscle biopsy could test for the presence of a particular synthetic gene or of a vector. But in the case of AAV, many people may be naturally infected with this harmless virus, so the test would not be conclusive for doping. Moreover, because most athletes would be unwilling to undergo an invasive biopsy before a competition, this type of genetic enhancement would remain virtually invisible.

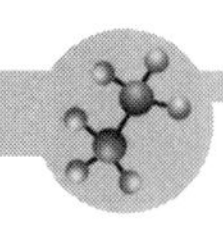

And what of the safety of rapidly increasing muscle mass by 20 to 40 percent? Could an athlete sporting genetically inflated musculature exert enough force to snap his or her own bones or tendons? Probably not. We worry more about building muscle in elderly patients with bones weakened by osteoporosis. In a healthy young person, muscle growth occurring over weeks or months would give supporting skeletal elements time to grow to meet their new demands.

This safety question, however, is just one of the many that need further study in animals before these treatments can even be considered for mere enhancement of healthy humans. Nevertheless, with gene therapy poised to finally become a viable medical treatment, gene doping cannot be far behind, and overall muscle enlargement is but one way that it could be used. In sports such as sprinting, tweaking genes to convert muscle fibers to the fast type might also be desirable. For a marathoner, boosting endurance might be paramount.

Muscle is most likely to be the first tissue subject to genetic enhancement, but others could eventually follow. For example, endurance is also affected by the amount of oxygen reaching muscles. Erythropoietin is a naturally occurring protein that spurs development of oxygen-carrying red blood cells. Its synthetic form, a drug called Epoietin, or simply EPO, was developed to treat anemia but has been widely abused by athletes—most publicly by cyclists in the 1998 Tour de France. An entire team was excluded from that race when their EPO use was uncovered, yet EPO abuse in sports continues.

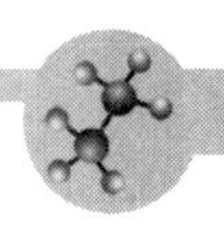

Gene transfer to raise erythropoietin production has already been tried in animals, with results that illustrate the potential dangers of prematurely attempting such enhancements in humans. In 1997 and 1998 scientists tried transferring synthetic erythropoietin genes into monkeys and baboons. In both experiments, the animals' red blood cell counts nearly doubled within 10 weeks, producing blood so thick that it had to be regularly diluted to keep their hearts from failing.

The technology necessary to abuse gene transfer is certainly not yet within reach of the average athlete. Still, officials in the athletic community fear that just as technically skilled individuals have turned to the manufacture and sale of so-called designer steroids, someday soon a market in genetic enhancement may emerge. Policing such abuse will be much harder than monitoring drug use, because detection will be difficult.

It is also likely, however, that in the decades to come, some of these gene therapies will be proved safe and will become available to the general population. If the time does arrive when genetic enhancement is widely used to improve quality of life, society's ethical stance on manipulating our genes will probably be much different than it is today. Sports authorities already acknowledge that muscle-regenerating therapies may be useful in helping athletes to recover from injuries.

So will we one day be engineering superathletes or simply bettering the health of the entire population with gene transfer? Even in its infancy, this technology clearly has tremendous potential to change both sports and our society. The ethical issues surrounding genetic enhancement are many and complex. But for once, we have time to discuss and debate them before the ability to use this power is upon us.

The rationale for large, multicelled, specialized organisms is based upon a fundamental biophysical principle. All cells rely on the process of diffusion across the cell membrane to bring nutrients into the cell and eliminate wastes from the cell. Diffusion requires a large surface area and a short distance to be effective. So, if we consider a spherical cell, the surface area of the cell is given by the formula ($A = 4\pi r^2$) and the volume is given by the formula ($V = 4/3\pi r^3$), where r is the radius of the sphere. As the radius doubles, the surface area and volume increase by factors of 4 and 8, respectively. Further increases in radius indicate that the volume (distance for diffusion) increases to a much greater degree than the surface area. So, as the cell increases in size, it is more difficult to transport substances by diffusion. To get around this basic biophysical limitation, large organisms evolved that are composed of large numbers of small cells with groups of those cells specialized to carry out various functions. In the next article, William P. Jacobs discusses one example of a

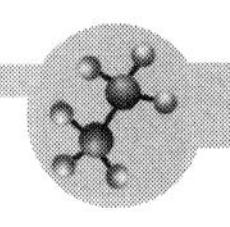

class of siphonous, or tubular, marine alga called Caulerpa, which is an apparent contradiction to this classic argument. —CCF

"*Caulerpa*"
by William P. Jacobs
***Scientific American*, December 1994**

This tropical alga is the world's largest single-celled organism. Yet it differentiates into a complex structure of leaves, stems and roots

Swimming with faceplate and snorkel over a lagoon where *Caulerpa* grows, you would be unlikely to notice anything unusual about this green plant protruding from the coral sand. It looks much like the sea grasses that also thrive in warm, shallow seas around the world. The horizontal stem and branched, leafy form of *Caulerpa* resemble those of many higher plants, such as the bracken fern or the strawberry plant. But an internal examination reveals the alga's uniqueness. *Caulerpa* is the largest, most differentiated single-celled organism in the world. No cell wall or membrane separates each of the many nuclei and their adjacent cytoplasm from the others.

This unusual, unexpected organism remains unknown to most biologists, although it was first described almost 150 years ago. By its very existence, *Caulerpa* is a gauntlet flung in the face of biological convention. No single cell should be capable of growing to a length of two or three feet, much less differentiating into separate organs, such as a stem, roots and leaves.

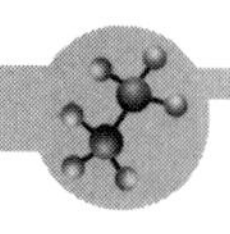

The questions raised by *Caulerpa*'s peculiar structure have intrigued the few biologists who have investigated its development.

All other organisms of such size and complexity consist of hundreds of thousands of microscopic cells. In each cell, a membrane encloses a limited volume of cytoplasm and a single nucleus. Most plants also secrete a cell wall outside the membrane. So ubiquitous is this organization that between 1838 and 1839 Matthias Schleiden of the University of Jena in Germany and Theodor Schwann of the University of Louvain in Belgium enshrined their observations in the form of the "cell theory." They posited that the cell is the basic unit of biological structure and function in both plants and animals. In the many decades since then, thousands of observations have converted the cell theory into a broadly accepted generalization that forms the basis of current ideas about biological development. Those ideas attribute the microscopic size of the average cell to the limited range of influence of the enclosed nucleus over its surrounding cytoplasm.

So how can *Caulerpa* grow to its macroscopic size and complexity without the compartmentalization that other organisms seem to need? Is *Caulerpa* handicapped by its single-celled form? Do hormones coordinate its development and regeneration, as they do in higher plants? If so, does *Caulerpa* employ the same chemical messengers as these plants do? How does *Caulerpa* keep all its cytoplasm from flowing out into the ocean when its only cell wall is breached by waves or hungry animals? I have spent the past 40 years looking for the answers to these questions.

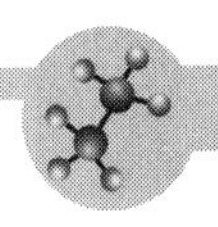

I became intrigued with *Caulerpa* in the early 1950s, after hearing of it from William "Cappy" Weston, a charismatic teacher at Harvard University. When I turned to the literature to learn more, I found an exasperating mess. A smattering of publications had appeared throughout the past century, each usually the result of a biologist's brief vacation visit to the Zoological Station in Naples, Italy. Contradictory observations made during these forays were impossible to resolve. Any of a number of factors could explain the often conflicting results. Seasonal changes from spring to fall or inadequate (and often unmentioned) sample size undoubtedly affected the findings. The visitors rarely had time to repeat any experiment. And, of course, many worked before the days of statistical analyses, leaving the reader to guess at the reality of differences reported.

Only after two European biologists decided to investigate *Caulerpa* over longer periods could one have some confidence in the published reports. J. M. Janse of the University of Leiden in the Netherlands made several summer visits to Naples between 1886 and 1909. Then, in the 1920s, Rudolph Dostál of the University of Brno in the Czech Republic took the alga back to his seawater aquarium in his laboratory and so was able to study it for longer intervals. In a 1945 paper he summarized his 20 years of research on regeneration of *Caulerpa* pieces. A decade later I began my *Caulerpa* investigations, starting with a six-month sojourn in Naples. Following Dostál's example, I brought the alga home and began growing it in a continuous culture in my laboratory at Princeton University.

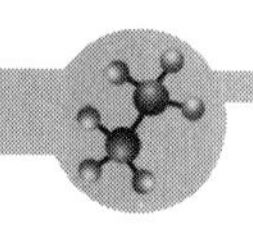

Working with my research associates, I have examined the development of *Caulerpa prolifera* in quantitative detail. Among our first discoveries was that the growth rate of the alga is comparable to that of multicellular organisms. James C. W. Chen, then at Princeton, and I took daily photographs to measure its growth. We found that the rhizome (the cylindrical stem of the plant) grew roughly 4.6 millimeters a day, a rate similar to that observed for stems of several multicellular plants.

The pattern of *Caulerpa*'s development, however, differs from the more complex growth seen in higher plants. Most multicellular plant organs mature at rates that vary with time, but *Caulerpa* elongates at a constant speed. In multicellular plants, some individual cells that contain multiple nuclei (as *Caulerpa*'s large cell does) also demonstrate this pattern. Thus, extended periods of constant growth may be typical of multinucleated cell structure, common both to *Caulerpa* and to certain cells in higher plants.

We were surprised to find that the elongation rates of all three organs of *Caulerpa*—stem, root and leaf—do not differ significantly from one another. In higher plants, the speed of growth varies from organ to organ. Our results from *Caulerpa* suggest that whatever factor limits development of the stem must pervade the entire plant so that it also limits the growth of the other two types of organs.

The localization of growth of *Caulerpa* stems also diverges from the patterns generally seen in organs of higher plants. We demonstrated that the *Caulerpa*

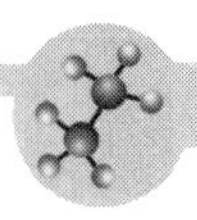

stem and roots extend only at their tips. The organs of multicellular plants, in contrast, show much more complicated patterns, rarely limiting elongation to that area.

My students and I next turned our attention to exploring the early development of *Caulerpa*'s root clusters and leaves. We found that they provided still more clues about the plant's unusual growth patterns. The regularly spaced root clusters, or rhizoids, result from the daily initiation of a new cluster along the bottom of the stem, close to its tip. Leaves develop on the upper side of the stem, farther back from its tip; they appear less often and with less regularity than do root clusters. In my cultured *C. prolifera* sample, collected from the Florida Keys, a new leaf formed every five or six days. The leaf initially resembled the pointed cylinder of a new root or stem, but the tip soon grew a flattened, heart-shaped blade and unfolded into a roughly rectangular form four to five inches long. Often another new leaf started growing near the top of the primary one once it had matured. In fact, the species name *prolifera* derives from the proliferation of leaves sprouting from the tops of older ones.

The leaves of *Caulerpa* shoot up from the top of the stem and extend toward the sunlight. The roots develop on the underside and grow down into the ocean floor. By analogy with multi-cellular plants, it seemed to me that these orientations might be controlled by directional signals from light or gravity, or both. So we set out to discover whether gravity alone was sufficient to control where leaves and roots emerge.

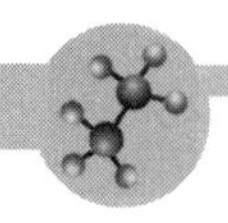

We turned one set of our *Caulerpa* plants upside down by rotating each plant 180 degrees around its horizontal axis. On the following day, we saw that the next root had developed on the new underside of the stem. It formed without any delay, as compared with the roots developing in the upright plants. The next leaf formed at its normal spacing but on the new upper side. This sequence of events held even if light came from both sides of the plants rather than from above as occurs in nature.

This is the fastest effect in altering the location of organ development that we know of. A local accumulation of the starch-storing structures inside cells known as amyloplasts apparently triggered the change. Michael B. Matilsky, then at Princeton, and I found that within six hours of inverting the plants there were 54 percent more amyloplasts on the new underside of the stem tip than before. We also noticed that the increased number of these organelles accumulated where the new root cluster would later develop. A corresponding decrease in the number of amyloplasts occurred on the upper side of the stem tip. Apparently the amyloplasts drifted down through the cytoplasm and settled to the bottom of the stem. Farther back from the tip, in the areas where roots do not develop, there was no redistribution of the organelles after inversion. Our results indicate that *Caulerpa* uses amyloplasts to respond to gravity just as higher plants do. Yet its reaction to gravity is somewhat different. Instead of changing only the direction of growth, gravity initiates the development of organs in novel locations on *Caulerpa*.

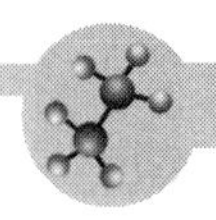

Any laboratory study of *Caulerpa* originally required the removal of some alga from the ocean. How does the single-celled plant withstand being torn open by grazing fish or sadistic biologists such as myself? As soon as a leaf or stem is cut, some cytoplasm does stream out into the seawater as one would expect. But a wound plug forms, and a new wall is laid down behind it, sealing off the remainder of the cell. Such self-sealing allows *Caulerpa* to survive substantial loss of leaf area and permits the process of regeneration to begin. Indeed, regeneration of an entire *Caulerpa* plant from a piece of leaf or stem is not uncommon.

Regenerating pieces of *Caulerpa* can often be found in the sea. For many years, the lack of any evidence for other forms of reproduction led biologists to consider regeneration as perhaps the only way the plant reproduced. But in the late 1920s Dostál observed that old leaves formed tiny projections from which cytoplasm oozed. Along with the cytoplasm, the plant released flagellated cells capable of movement. By the end of the 1930s several people had observed sexual reproduction—the fusion of pairs of such cells—in different *Caulerpa* species. Nevertheless, tearing and subsequent regeneration still appear to be the most probable way that *Caulerpa* reproduces in nature.

An experimental version of a wound plug permits closer study of *Caulerpa*'s regeneration mechanisms. In 1904 Janse discovered that pressing together opposite walls of the cell and clamping them would lead in a few days to the development of a new cell wall known as a pressure wall. More recently I have found that this protective response happens much faster, on the order of minutes. We can then cut the alga apart at pressure walls with little loss of cytoplasm and observe the regeneration of small pieces of *Caulerpa*. Such pieces cannot regenerate otherwise, because there is not enough cytoplasm remaining to reorganize the cellular material and trigger growth.

Regeneration of the whole plant from large pieces occurs naturally; laboratory studies indicate an unexpected directionality to the process. For example, when we cut a 50-millimeter-long piece from a *Caulerpa* leaf, an entire plant regenerated in a particular polar sequence. First, roots developed at the cut made closer to the stem. Later, a new stem grew only a few millimeters away from this cut. On the outer half of the leaf section, a new leaf began to regenerate. Most organisms exhibit such polarity in normal development as well as in regeneration (if they are capable of that). But in multicellular plants, such regeneration is usually attributed to the polar movement of growth substances through thousands of cells. Demonstration of a similar trait in single-celled *Caulerpa* surprised many.

We also used pressure walls to alter developmental pathways in regenerating pieces of *Caulerpa*. A transverse pressure wall made just below a growing leaf places the bud physiologically above the base of the section. Because the sprout sits near the bottom of its stem, the little nubbin that would have grown out as a leaf will adjust to become a root. Similarly, a bud starting as a root can often be induced to change to a leaf instead.

In view of the striking effect of gravity on the development of the stem tip, we also wondered how gravity might affect regeneration. Because leaves also contain the starch-storing amyloplasts, perhaps their gravity-induced resettling could alter regeneration patterns as well. But the pattern and number of regenerated organs were unchanged by inverting the leaf pieces. For instance, roots regenerated exclusively on the original stem end of the leaf piece, whether that end was up or down relative to gravity.

It is still unclear what determines the directionality of regeneration of pieces of *Caulerpa*. The most likely candidate is the striking cytoplasmic streaming. Strands of moving cellular material are easily visible in the leaves, even under the low magnification of a dissecting microscope. The broad channels run roughly parallel to the long axis of the leaf, and the direction of the flow can be opposite in adjacent strands. The streams may move organ-forming substances in a particular direction, thus substituting for the myriad direction-specific transporting cells of higher plants.

If one ignored the fact that these regenerating fragments are simply multinucleated pieces of a single cell, *Caulerpa*'s polar regeneration seems much like the well-known regeneration of pieces of other plants. Do *Caulerpa* species—and by extension do other algae—use hormones to coordinate development and regeneration? If so, are the hormones chemically related to those used by presumably more highly evolved plants?

The hormone indole-3-acetic acid (IAA) helps to control regeneration in multicellular plants. We have recently documented conclusively, using gas chromatography and mass spectrometry, that IAA is present in *Caulerpa*. Work done earlier by Ned Kefford and Arun Mishra of the University of Hawaii, Clinton J. Dawes of the University of South Florida and Henry Augier of the University of Aix-Marseille II in France showed that addition of IAA to the seawater in which *Caulerpa* thrives stimulates its growth patterns. But even with evidence that IAA is present and active in *Caulerpa*, we still wondered if the action of IAA depended on selective distribution of the hormone, as it does in multicellular plants. No evidence exists for the formation of pockets of IAA within *Caulerpa*. To the contrary, if IAA labeled with radioactive carbon is added to the tip of a *Caulerpa* leaf, the radioactivity—and presumably IAA—spreads uniformly along the stem.

Gibberellins are another class of growth hormones active in many multicellular plants. Early studies reported that extracts of *Caulerpa*, when added to higher plants, triggered growth patterns just as pure gibberellins do. But my recent, more extensive analysis using mass spectrometry revealed no known gibberellin or gibberellin metabolite in such *Caulerpa* extracts. Was the gibberellinlike activity seen in *Caulerpa* caused by a chemical that did not have the basic structure of the compound but that happened to show similar activity? Or was the activity from yet another, as yet unknown, gibberellin that could be added to the already long list of more than 70 of these compounds?

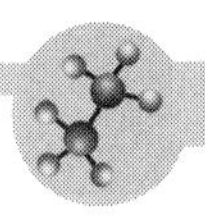

We surmise that hormones produce their effects in *Caulerpa* by interaction with substances or organelles whose distribution does vary. For example, the amyloplasts that accumulate at the bottom of the growing stem tip may promote root formation by working in concert with either IAA or the gibberellinlike substance. The interaction maintains gravity-directed growth and initiates root formation at specific sites.

Although the details of many of *Caulerpa*'s hormonal interactions are unclear, most of its organelles are known and resemble those of higher plants. In the peripheral cytoplasm, chloroplasts used in photosynthesis and many small nuclei stream along with the starch-storing amyloplasts. A large cytoplasmic sac, or vacuole, of convoluted shape sits in the center of the cell.

The only visibly unusual features are numerous rods that project in from the cell wall that encloses the whole plant. The rods run both perpendicularly and parallel to the long axes of the organs, forming a dense, interconnecting network. Janse counted 850 rods per square millimeter in the older part of the leaf and five times as many in the tip. The rods are sheathed in cytoplasm, and their density throughout the organism partially compensates for the lack of cell membrane surface. They appear to serve as a supporting skeleton for the huge cell. A few investigators have suggested that they also serve as conduits to and from the surrounding sea.

Dinkar D. Sabnis, working with me at Princeton, looked at *Caulerpa* with an electron microscope and found a similarly unusual distribution of tiny fibers

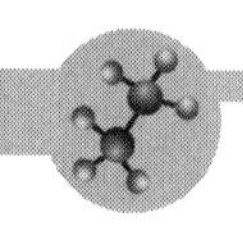

known as microtubules. We determined that sheets or thick bundles of microtubules, evenly spaced and uniformly oriented, were arrayed in the internal layer of cytoplasm where streaming occurs. Our suggestion that microtubules were related in some way to streaming was later confirmed by other researchers. Inoculating the plant with the alkaloid compound colchicine disassembled the microtubules (as expected from research on various multicellular organisms) and stopped the streaming.

Many of the riddles posed by *Caulerpa*'s existence are now understood. Despite its uniquely large, single-celled organization, *Caulerpa* competes successfully with multicellular organisms that inhabit the world's warmer seas. It grows as fast as many of its land-based, multicellular, distant relatives. *Caulerpa* has proved so hardy that one species is raised commercially in sea-water pens in the Philippines for use in fresh green salads. It readily regenerates entire plants from pieces of stem or leaf and does so in a temporal and spatial pattern resembling that seen in more highly evolved, multicellular plants. The fact that 73 species of *Caulerpa* exist around the world, making them far from rare in the tropical algal flora, suggests that their single-celled construction is not a great handicap.

Musing over the current knowledge we have about *Caulerpa* and the many questions still remaining, I find that evolution has provided more possibilities than we have tended to expect. If *Caulerpa* is this prominent as a large, highly differentiated, multinucleated single cell, what are the ultimate lengths to which this structure could be carried? I can see nothing that would preclude an even larger unicellular organism so long as it lives in the sea. There the buoyant water substitutes for the internal support provided to land plants by their cell walls. Might we one day discover a huge algal equivalent of "Audrey II," the outrageously large Venus flytrap of *The Little Shop of Horrors*?

2 Differentiation: The Path to Specialization

As mentioned in the example of the human embryo in the introduction, development involves the processes of cell division, changes in cell shape, cell migration, differentiation, and apoptosis. All of these processes in turn depend upon various biochemical processes, such as the coordination of gene expression and signals from one cell to another, such as induction. Therefore, to study these processes, both cellular and biochemical, scientists need various model systems. Ideally, these systems must exhibit all of the processes necessary for development and differentiation and must occur over a short time period. Another important aspect of any model system is that mutations (changes in the DNA, either naturally occurring or artificially induced) must be found. Such mutations allow scientists to identify genes that are important to the process and to assess their relative importance in differentiation. To this end, scientists have used fruit flies, worms, frogs, zebrafish, and plants as model systems. In the article here, Martin

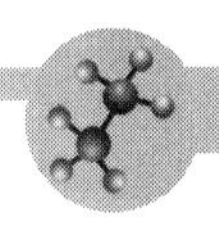

Hülskamp discusses how leaf hairs, or trichomes, can be used as a model to study all of the cellular processes—cell fate, control of the cell division cycle, cell polarity, and changes in cell shape—that occur in plant differentiation. —CCF

"Plant Trichomes: A Model for Cell Differentiation"
by Martin Hülskamp
***Nature Reviews Molecular Cell Biology*, June 2004**

During the development of multicellular organisms many cell types are produced. Depending on their position, each cell perceives different signals, responds through intracellular signalling pathways and, eventually, adopts a specific cell fate. Subsequent cell differentiation usually involves complex changes. For example, cells exit the mitotic cycle or enter cycles of ENDOREDUPLICATION, the cellular architecture alters to meet the functional requirements of the respective cell type, and the metabolism of the cell changes according to its function. Compared with animals, plant development faces additional constraints because the rigid cell walls prevent any cell movement. In plants, a few single-celled *Arabidopsis thaliana* model systems—in particular root hairs and trichomes—have greatly improved our understanding of the development of single cells.

The goal of this review is to summarize how the study of *A. thaliana* trichomes facilitates the understanding of development at the single-cell level. A large number of mutants have been characterized that enabled the identification of subsequent developmental processes.

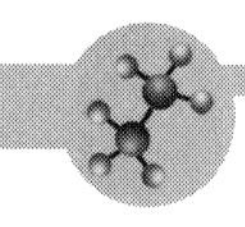

These include the selection of trichomes in a field of epidermal cells, cell-fate determination, changes in the cell-cycle mode and cell-shape control. The genetic, molecular and cell-biological analysis of trichome development has revealed only a few trichome-specific processes, as most developmental steps involve the regulation of general cellular machineries. Therefore, studying the trichome system has provided unique insights into the function of transcription factors, the microtubule and actin cytoskeleton, the cell cycle and cell-death control. The study of all the developmental stages of a single cell is a first step towards an understanding of how general cellular processes are integrated during development.

Steps in Trichome Development

Shoot epidermal hairs are known as trichomes, a term that is derived from the Greek word for hairs, *trichos*. Trichomes are found in most plants and can comprise either single or several cells and can be secretory glandular or nonglandular.[1,2] The functions that are ascribed to trichomes range from protecting the plant against insect herbivores and UV light, to reducing transpiration and increasing tolerance to freezing.[3,4]

Trichomes are an excellent model system because they are of epidermal origin and are therefore easily accessible. In addition, *A. thaliana* trichomes are not essential for the plant under laboratory conditions, which facilitates the isolation of trichome-specific mutants.[5,6] So far, most studies have been carried out on leaf trichomes. At the base of young leaves, single cells that are spaced out at regular distances in an area of

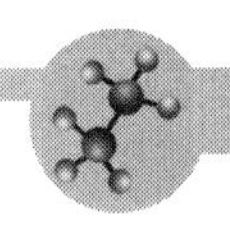

apparently equivalent PROTODERMAL CELLS develop into trichomes.[7, 8] Incipient trichomes stop mitotic cell divisions and initiate endoreduplication cycles. As a result, the trichome cell increases in size and changes its direction of growth such that it grows perpendicular to the leaf surface. Further growth is characterized by a total of about four endoreduplication cycles that result in a DNA content of 32C (1C is the DNA content of the unreplicated haploid genome), which is accompanied by rapid cell enlargement.[7, 9] The growing cell undergoes two consecutive branching events, the orientations of which are co-aligned with respect to the basal-distal leaf axis.[10]

A large number of mutations that affect trichome development were identified in several mutagenesis screens.[5, 6] The mutations helped to define regulatory processes in trichome development according to their specific developmental defects. The selection of trichome cells and the initiation of the trichome cell fate are under the control of a small group of so-called patterning genes. One gene seems to specifically translate the patterning cues into cell-fate differentiation. The switch from mitotic cycles to endoreduplication cycles and the number of endoreduplication cycles are controlled by the endoreduplication genes. A large number of genes are known to be important for branching. The directionality of expansion growth is affected in the so-called *distorted* mutants. One mutant is known to cause unscheduled cell death, and several other mutants seem to affect the maturation of the trichome.

Now that the genetic interactions are well understood and most of the genes have been cloned, the emerging picture is that only very few genes are, in fact, trichome specific. Most genes are relevant for many cell types and are involved in general cellular processes. It seems that mutations in these genes have little effect in most cell types but are crucial during trichome development—possibly because trichomes, with their rapid growth and enormous size, are more demanding.

Trichome Patterning and Initiation

Wild-type trichomes are initiated with an average distance of about three cells between developing trichomes, and almost never form directly next to each other—as would be expected if they were randomly distributed—which indicates that there must be an underlying patterning mechanism.[7, 8] A mechanism that would explain trichome patterning by a standardized cell division pattern that segregates trichome cell fates was excluded by clonal analysis.[8, 11] It is therefore hypothesized that trichome selection is based on a mutual inhibition mechanism:[12–15] cells that are initially equivalent produce a trichome-promoting factor (or factors) that activates a factor (or factors) that suppresses trichome development in the neighbouring cells.This way, cells begin to compete and, due to stochastic fluctuations, individual cells will gain higher levels of the promoting factor, produce more of the suppressing factor and, in turn, inhibit the neighbouring cells more strongly. Eventually, these cells would become committed to the trichome cell fate. For this mechanism to work a number of criteria have to be met. First, the positive and the negative regulators need to be involved in a feedback loop with the activator activating

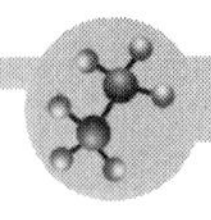

the inhibitor and the inhibitor inhibiting the activator. A second requirement is that the inhibitor can move.

The genetic and molecular analysis of the trichome patterning genes is consistent with this model, although little proof is available that could directly demonstrate the patterning mechanism. Both the positive and the negative regulator are represented by a group of several factors. Four of the trichome-patterning genes function as positive regulators of trichome development. Mutations in the *GLABRA1* (*GL1*) and *TRANSPARENT TESTA GLABRA1* (*TTG1*) genes each result in the complete absence of trichomes,[16, 17] whereas *GLABRA3* (*GL3*) and *ENHANCER OF GL3* (*EGL3*) function in a redundant manner—*gl3* mutants exhibit fewer trichomes compared with wild-type plants, whereas *gl3 egl3* double mutants are devoid of trichomes.[18] The trichome-suppressing genes are represented by three redundantly acting genes. Mutations in the *TRIPTYCHON* (*TRY*) gene result in trichome clusters, mutations in the *CAPRICE* (*CPC*) gene cause an increased number of trichomes[19, 20] and a single mutant of *ENHANCER CAPRICE TRIPTYCHON1* (*ETC1*), which is an enhancer of *cpc* and *try* mutants, is indistinguishable from wild-type plants.[21]

Genetic analysis has established the functional relationships between the four positive factors. The findings that co-overexpression of GL3 and EGL3, as well as GL3 together with GL1, can rescue the *ttg1*-mutant phenotype indicates that TTG1 functions upstream of these genes and that the other three factors function together at the same point in the pathway.[18, 22] These data have been confirmed at the molecular level. All trichome-promoting genes, except for *TTG1*, encode putative transcription factors. *GL1* encodes a MYB-RELATED TRANSCRIPTION FACTOR,[23] *GL3* a BASIC HELIX–LOOP–HELIX (BHLH) FACTOR,[22] *EGL3* is a close homologue of *GL3* (REF. 18), whereas *TTG1* encodes a WD40 PROTEIN whose molecular function is unknown.[24] Yeast two-hybrid data indicate that the four positive factors form a transcriptional-activation complex in which GL3 forms a homodimer that binds to GL1. (REF. 18, 22) The GL3 protein also binds to the TTG1 protein, but through a different domain. No direct interaction was found between GL1 and TTG1. (REF. 22) It is likely that GL1 and GL3 mediate the transcriptional activation, as both proteins contain transcriptional-activation domains. This complex is expected to be active in trichome precursor cells and be inactivated in all other protodermis cells by one or more known negative regulators of trichome initiation.

The negative regulators TRY, CPC and ETC1 all belong to a small family of single-repeat MYB proteins with no obvious transcriptional-activation domain.[19–21] Overexpression of any of these proteins abolishes trichome formation. They seem to function in a highly redundant manner: *try* mutants have small trichome clusters consisting of two or three trichomes, *try cpc* double mutants have large clusters of up to 40 trichomes and in *try cpc etc1* triple mutants, fields of several hundred trichomes are observed.[7, 19, 21] These negative factors seem to interfere with the function of the transcriptional-activation complex by a competition mechanism.[25] Three-hybrid analysis has shown that the interaction between GL1 and GL3 is counteracted by TRY, thereby

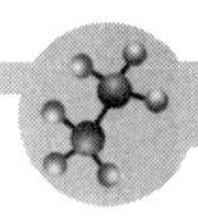

disturbing the formation of the proposed functional transcriptional-activation complex.[26]

How cell–cell communication and a regulatory feedback loop are achieved is, at present, unknown. However, evidence is available for root-hair patterning in *A. thaliana*, which requires a set of identical and closely related genes. Here it was shown—using a green fluorescent protein (GFP)–CPC fusion protein—that the negative regulator CPC moves between cells, probably via PLASMODESMATA, which indicates that travelling transcription factors can mediate cell–cell communication.[27] It was also shown that the positive root-hair-patterning genes *WEREWOLF* (*WER; a GL1 homologue*) and *GLABRA2* (*GL2*) are involved in a negative feedback loop with *CPC*.[28] It is conceivable that a similar mechanism operates during trichome patterning.

Trichome Differentiation

The homeodomain leucine-zipper protein that is encoded by the *GL2* gene[29, 30] is thought to translate the cues that are provided by the patterning genes into cell-specific differentiation of several epidermal cell types including the seed coat, root hairs and trichomes.[17, 29, 30] This is documented by the finding that cooverexpression of GL1 and the maize *R* gene product (a homologue of GL3) results in an increased and ectopic expression of GL2. (REF. 31) Although some evidence indicates that GL2 might also have a role in trichome patterning, most of the available data indicate that GL2 triggers downstream differentiation events.[32] The *gl2*-mutant-trichome phenotype is characterized by

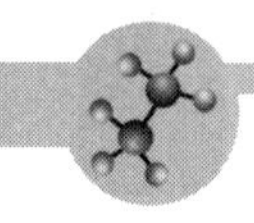

undifferentiated trichomes that resemble the combined phenotypes of various trichome-morphogenesis genes, which indicates that GL2 activates trichome-specific differentiation genes.[7, 29] Supporting evidence comes from the root-hair system, where it was shown that GL2 regulates the gene that encodes phospholipase Dζ1, which, in turn, promotes root-hair differentiation.[33]

Cell-Cycle Control During Trichome Development

Trichome cell-cycle-regulation mutants affect either the switch from mitotic cycles to endoreduplication cycles, or the number of endoreduplication rounds, and thereby the ploidy level.

The *SIAMESE* (*SIM*) gene suppresses the switch from mitotic divisions to endoreduplication cycles.[34] In *sim* mutants, trichomes are multicellular and contain between 2 and 15 cells. If the first cycle is already mitotic, two trichomes are formed instead of one; if the switch to endoreduplication from mitotic divisions is late, multicellular trichomes that are morphologically normal are formed. As *SIM* has not been cloned yet, little is known about the molecular mechanism that is involved. However, some information is available from a different line of experiments in which the role of known cell-cycle genes in the control of endoreduplication was tested. The expression of cell-cycle genes that are normally not active in trichomes was used to test the effects on trichome endoreduplication. To avoid organism-wide defects that could cause sickness or even lethality, trichome-specific gene promoters were used. B-TYPE AND D-TYPE CYCLINS could trigger the formation of multicellular trichomes, and B-type cyclins are involved in the transition from G2 phase to mitosis. The overexpression of a specific B-type cyclin, CYCB1;2, is sufficient to switch from endoreduplication cycles to mitotic cycles, which indicates that B-type cyclins are important for this regulatory step. Surprisingly, the overexpression of the D-type cyclin CYCD3;1 can also lead to the switch.[36] D-type cyclins are thought to control the transition from the G1 to the S phase of the cell cycle in animals; but these results indicate that, in plants, they have an additional function in regulating the entry into mitosis. In *sim* mutants, CYCB1;2 is expressed in trichomes, which indicates that SIM inhibits the expression of mitotic cyclins. However, this is not its only function, as overexpression of CYCB1;2 in a *sim* mutant background shows a much stronger phenotype than the single mutants. By contrast, the initiation of mitotic cycles in *sim* mutants is independent of D-type cyclins.[36]

The number of trichome endoreduplication cycles is affected in at least ten mutants that display either lower or higher ploidy levels than normal. The picture that is emerging from their genetic and molecular analysis is that several different molecular pathways are involved in the regulation of trichome endoreduplication cycles.

Regulation by patterning genes. Two of the patterning genes that are described above, *GL3* and *TRY*, also function as positive and negative regulators of endoreduplication cycles; *try* trichomes have a DNA content of 64C and different *gl3* alleles exist that have either a reduced or an increased DNA content.[7, 26] This dual function raises the fascinating possibility that trichome cell-fate choice is functionally linked with cell-cycle regulation.

DNA-catenation-dependent endoreduplication. ROOT HAIRLESS2 (RHL2) and HYPOCOTYL6 (HYP6) are positive regulators of endoreduplication cycles in trichomes and in other cell types.[37] Both are plant homologues of the archaeal DNA TOPOISOMERASE VI complex. These topoisomerases can DECATENATE DNA and promote ATP-dependent separation of entangled DNA.[37] It is unclear whether the observed defect in the progression of trichome endoreduplication is due to a physical block of further DNA replication or the activation of a checkpoint that controls progression of the endoreduplication cycle.

Regulation by plant hormones. The plant hormone gibberellin promotes endoreduplication cycles. In *spindly* (*spy*) mutants, which exhibit a constitutive gibberellin response, trichomes have a DNA content of 64C. (REF. 38) Conversely, in a mutant that is incapable of gibberellin synthesis, *ga1-3*, no trichomes are formed.[39, 40]

Regulation by protein degradation. A class of four trichome mutants, *kaktus* (*kak*), *rastafari* (*rfi*), *polychome* (*poc*) and *hirsute* (*hir*), all show a very similar phenotype: they all have a ploidy level of 64C. The cloning of the *KAK* gene revealed that it encodes a protein with sequence similarity to a UBIQUITIN E3 LIGASE.[41, 42] It is

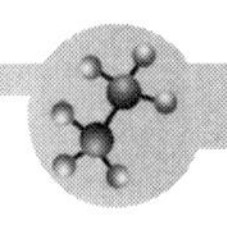

therefore assumed that ubiquitin-regulated protein degradation controls the progression of endoreduplication.

Regulation by a cell-death pathway. Two lines of evidence indicate that a pathway exists that controls both the progression of endoreduplication cycles (and mitotic cycles) as well as programmed cell death. First, the *CONSTITUTIVE PATHOGEN RESPONSE5* (*CPR5*) gene is involved in both processes. Second, overexpression of an inhibitor of the cell-cycle kinase *INHIBITOR/INTERACTOR OF CYCLIN-DEPENDENT KINASES/KIP-RELATED PROTEINS* (ICK/KRP) leads to reduced ploidy and early trichome cell death.

Trichome Branching

The typical three-dimensional branching pattern of trichomes is a unique model system for studying how several axes of polarity and cell morphogenesis are established. Except for the *sim* mutant, all mutants described so far affect the number of branches but not their orientation with respect to each other. Genetic and molecular data indicate that several independent molecular pathways participate in trichome branching.[7, 10, 43–46]

Regulation by endoreduplication levels. The number of branches on a trichome depends on the ploidy level of the cell. Tetraploid plants, in which the DNA content of all cells is doubled, have trichomes with supernumerary branches.[47] Similarly, mutants that have trichomes with increased DNA levels—such as *kak*, *poc*, *rfi*, *try* and *spy*—have trichomes with up to eight branches.[7, 47] The reduction of the ploidy level results in a reduced branch number—as found in *cpr5*, *gl3 rhl2* and *hyp6*. (REFS. 7, 37, 48) It is likely that this observed control of trichome branching by the ploidy level is indirect, and that the cell size or the time of actual cell growth provides the frame for branch initiation.

Regulation by microtubules. Microtubules have an important role in trichome branching, as shown in experiments with microtubule antagonists. If the microtubule cytoskeleton is defective during trichome growth, the cell expands almost equally in all directions (this is known as isotropic growth) and does not initiate branches.[49] Several branching genes encode components that are involved in the biogenesis of α/β-tubulin dimers, the formation and stability of microtubules or microtubule-based transport processes. Two weak mutants of *TUBULIN FOLDING COFACTOR* (*TFC*)*A* and *TFCC*—which are involved in the correct folding of tubulins and therefore the formation of assembly-competent α/β-tubulin dimers—exhibit a "bloated" and underbranched trichome phenotype.[50–52] The analysis of the microtubule cytoskeleton in these mutants sheds some light on how the microtubules are reoriented during branch initiation. As the microtubule density and orientation is normal, it is likely that the failure of branch formation is due to problems in *de novo* synthesis rather than in the reorientation of pre-existing microtubules. If branching requires the synthesis of new microtubules, it is conceivable that it also requires

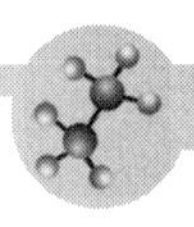

the fragmentation of pre-existing ones to allow a reorientation of growth. This view is supported by the reduced-branching phenotype of mutants of the *KATANIN-P60* gene.[53–55] Katanins are known to cut preexisting microtubules into smaller fragments. Microtubule reorientation during branch formation is therefore assumed to be controlled by severing preexisting microtubules combined with *de novo* synthesis.

The spatial control of the orientation of microtubules is regulated by at least two branching genes. The *FASS/TONNEAU2* gene regulates microtubules in the context of cell divisions as can be inferred from the observation that in *fass/tonneau2* mutants cell-division orientation is randomized.[56, 57] It encodes a novel protein, phosphatase-2A regulatory subunit, which indicates that it regulates microtubules by protein phosphorylation.[58] The second regulator of microtubule organisation is SPIKE.[59] This protein shows sequence similarity to CDM-family adaptor proteins (*Caenorhabditis elegans* CED-5; *Homo sapiens* DOCK180; *Drosophila melanogaster* MYOBLAST CITY). These proteins function as guanine nucleotide-exchange factors (GEFs)[60] and are thought to modulate the cytoskeleton through small RHO-like GTPases (known as ROPs in plants).[61]

In addition, specific microtubule-based transport processes seem to be important for branch formation. The branching gene *ZWICHEL* (*ZWI*) encodes a calmodulin-binding kinesin motor protein that binds microtubules in a calmodulin-dependent manner.[62–67] The activity of ZWI is modulated by the KIC protein, which binds to ZWI in a Ca^{2+}-dependent manner.[68] This indicates that ZWI-dependent transport processes might ultimately be controlled by the intracellular second messenger Ca^{2+}.

Regulation by transcription or Golgi-related processes. The *ANGUSTIFOLIA* (*AN*) gene regulates branching by two possible pathways, by Golgi-related transport processes or by transcriptional co-activation. It encodes a protein with sequence similarity to carboxyterminal binding protein (CtBP) and brefeldin-A-ribosylated substrate (BARS).[69, 70] In *D. melanogaster*, CtBP binds to the zinc-finger transcription factors and functions as a transcriptional co-repressor.[71] In the rat, BARS proteins were identified as proteins that are ADP-ribosylated after treatment with the fungal toxin brefeldin A. Brefeldin-A treatments result in the transformation of Golgi stacks into a tubular-reticular network and it is therefore thought that BARS is involved in Golgi functions.[72,73] Biochemical data are not available for the plant CtBP/BARS protein; however, the findings that *an* mutants have microtubule defects and that AN physically interacts with ZWI in a yeast two hybrid screen indicates that AN regulates microtubule organization.[69]

Regulation by the* STICHEL *gene. The *STICHEL* (*STI*) gene regulates trichome branching in a dosage-dependent manner; branch reduction is subtle in weak *sti* alleles, becomes more pronounced in stronger alleles and trichomes are unbranched in null-alleles. Conversely, overexpression of STI leads to extra branch formation.[74] This genetic behaviour indicates a key regulatory role for

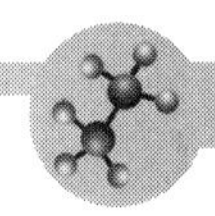

STI, although its molecular function is still elusive. *STI* encodes a protein that contains a domain with sequence similarity to eubacterial DNA-polymerase-III subunits. However, it is unlikely that STI functions as a DNA polymerase subunit, as no replication effects were found to be associated with the branching phenotype.

An underlying scheme of how branch formation is controlled is not evident from the current analysis of the branching genes. One model, however, accommodates all the available data by assuming that branching is evolutionarily derived from multicellular trichomes, in which branching is the result of a certain division pattern.

Directionality of Trichome Cell Expansion

Like most plant cells, trichomes enlarge several-fold during the later stages of differentiation and expand in a polarized manner. This expansion occurs, unlike in the growing tip of root hairs or pollen tubes, along the whole cell surface.[75, 76] The directionality of expansion growth is affected in mutants of eight genes, which are collectively known as the *DISTORTED* genes. All *distorted* mutants show a very similar phenotype: trichomes show turns and twists, some regions of the cell become bulged and others are underdeveloped. Following the movement of small beads that had been placed on the trichome surface, it was shown that this phenotype is caused by the regionally unbalanced expansion of the cell.[76]

Findings from different experimental approaches indicate that the directionality of trichome cell expansion depends on the actin cytoskeleton. First, the application of drugs that interfere with actin function perfectly phenocopies the *distorted* mutant phenotype.[75, 77] Second, the actin cytoskeleton is organized aberrantly in *distorted* mutants.[75–77] Third, all *DISTORTED* genes that have been cloned so far encode components of the ARP2/3 COMPLEX,[78–81] which promotes actin polymerization by enhancing F-actin nucleation and side-binding activities that result in the initiation of fine actin filaments from pre-existing F-actin.[82, 83]

The analysis of the *distorted* mutants demonstrated that actin has a role in expansion growth that goes beyond its mere requirement for general growth. The observation that in *distorted* mutants actin-based movement of organelles, such as peroxisomes or the Golgi, is not generally impaired indicates that F-actin is still functional.[78, 79] Defects were found locally in those parts of the cell that were not growing. Nongrowth regions contain heavily bundled actin, whereas regions in the distorted mutants that exhibit growth comprise a fine network of actin known as "fine actin." It is conceivable that the creation of a local fine-actin network promotes the transport of membrane and cell wall material for the actual growth. It is speculated that the actin cytoskeleton is also involved in the fusion of membranes, as the fusion of small vacuoles, which normally leads to the formation of the large central vacuole, does not take place in *distorted* mutants.[79]

It is unknown how the ARP2/3-complex-dependent formation of fine actin is spatially controlled in trichomes. Some of the canonical pathways such as the RHO and RAC/CDC42 signal-transduction pathways that are known in animals and yeast are, in principle, present in plants, although they are strongly modified. In

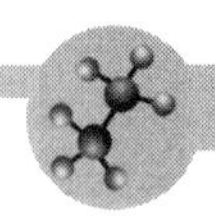

agreement with this, ROPs were shown to control the local actin configuration in epidermal cells and downstream components, such as the HSPC300 (haematopoetic stem/progenitor-cell clone-300) complex, are known to be involved in the control of actin organization.[84–86]

Cell-Death Control in Trichomes

The analysis of trichome development has revealed two pathways that suppress cell death and also regulate endoreduplication (see above). One pathway is represented by ICK/KRP, which shows homology to the animal cell-cycle inhibitor $p27^{Kip1}$ (REF. 87). In animals, p^{27Kip1} can induce apoptosis in the absence of growth factors in some specific cell types.[88] When ubiquitously expressed in the whole plant, ICK/KRP causes severe growth reduction,[87, 89, 90] and when expressed under the control of a trichome-specific promoter, trichome cells stop endoreduplication cycles after two cycles and begin to die with symptoms that are characteristic of programmed cell death, such as the degeneration of CHROMOCENTRES and nucleoli.[91]

A second pathway is linked to the response of plants to plant pathogens. A number of mutants mimic the plant pathogen response. Many of these mutants show a cell-death phenotype combined with growth defects.[92] One of these mutants, *cpr5*, shows a trichome phenotype that is similar to that of ICK/KRP-overexpressing lines; trichomes have a ploidy level of about 8C and undergo unscheduled cell death.[48] It seems that in both cases cell-cycle or endoreduplication-cycle progression and the control of

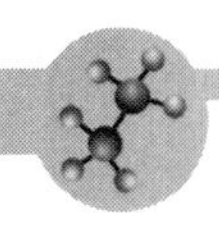

cell death are somehow linked; however, the mechanistic basis of this link remains to be determined.

Control of Maturation

Trichome maturation is affected in a group of diverse mutants in which adult trichomes appear transparent or underdeveloped. Three poorly characterized mutants, *chablis*, *chardonnay* and *retsina*, have transparent trichomes and the underdeveloped trichome mutant has no papilla on the trichome surface.[93] The trichome *birefringence* mutant is defective in the production of cellulose.[94]

Conclusions and Perspectives

Almost all trichome genes are involved not only in trichome development, but also in the development of other cell types and represent important components of generally important regulatory pathways. The analysis of trichome initiation has uncovered an evolutionarily conserved gene cassette of transcription factors that are involved in patterning processes and anthocyanin-synthesis control. Their evolution and functional diversification will be very interesting to study.

Also, the theoretical model that explains pattern formation is far from being proven; for example, at present, there are no target promoters known that could be used to test the genetic predictions. Therefore, it will be challenging to show not only that the inhibitor proteins can move, but also how this is relevant for patterning.

Several pathways seem to have a role in how the switch from mitosis to endoreduplication and the cycle number are controlled. The isolation of further genes, in

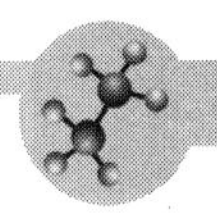

combination with trichome-specific overexpression approaches, should be a valuable addition to the cell-cycle field. The analysis of branching genes has led to the identification of proteins that are involved in processes as different as intracellular transport, cell-size control, transcriptional control and Golgi-dependent processes, as well as still unknown processes such as those controlled by STI. Each group of genes has opened new research areas in the plant sciences and it will be interesting to see whether the common branching phenotype will tie these processes together. With the discovery that cell-expansion genes encode components of the ARP2/3 complex, key components that regulate actin-based growth have been identified and will allow the study of the up- and downstream regulatory processes in plants. Further analysis of trichomes as a single-cell model system offers the chance to connect the above-mentioned, seemingly unrelated, processes in the future.

REFERENCES

1. Esau, K. *Anatomy of Seed Plants* (John Wiley & Sons, New York, 1977).
2. Uphof, J. C. T. *Plant hairs* (eds. Zimmermann, W. & Ozenda, P. G.) (Gebr. Bornträger, Berlin, 1962).
3. Johnson, H. B. Plant pubescence: an ecological perspective. *Bot. Rev.* 41, 233–258 (1975).
4. Mauricio, R. & Rausher, M. D. Experimental manipulation of putative selective agents provides evidence for the role of natural enemies in the evolution of plant defense. *Evolution* 51, 1435–1444 (1997).
5. Marks, M. D. Molecular genetic analysis of trichome development in *Arabidopsis. Annu. Rev. Plant Physiol. Plant Mol. Biol.* 48, 137–163 (1997).
6. Hülskamp, M., Schnittger, A. & Folkers, U. Pattern formation and cell differentiation: trichomes in *Arabidopsis* as a genetic model system. *Int. Rev. Cytol.* 186, 147–178 (1999).
7. Hülskamp, M., Misera, S. & Jürgens, G. Genetic dissection of trichome cell development in *Arabidopsis. Cell* 76, 555–566 (1994).

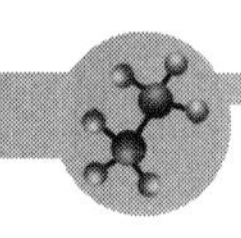

8. Larkin, J. C., Young, N., Prigge, M. & Marks, M. D. The control of trichome spacing and number in *Arabidopsis. Development* 122, 997–1005 (1996).
9. Melaragno, J. E., Mehrotra, B. & Coleman, A. W. Relationship between endopolyploidy and cell size in epidermal tissue of *Arabidopsis. Plant Cell* 5, 1661–1668 (1993).
10. Folkers, U., Berger, J. & Hülskamp, M. Cell morphogenesis of trichomes in *Arabidopsis*: differential control of primary and secondary branching by branch initiation regulators and cell growth. *Development* 124, 3779–3786 (1997).
11. Schnittger, A., Folkers, U., Schwab, B., Jürgens, G. & Hülskamp, M. Generation of a spacing pattern: the role of *TRIPTYCHON* in trichome patterning in *Arabidopsis. Plant Cell* 11, 1105–1116 (1999).
12. Hülskamp, M. & Schnittger, A. Spatial regulation of trichome formation in *Arabidopsis thaliana. Semin. Cell Dev. Biol.* 9, 213–220 (1998).
13. Scheres, B. Plant patterning: TRY to inhibit your neighbors. *Curr. Biol.* 12, R804–R806 (2002).
14. Schiefelbein, J. Cell-fate specification in the epidermis: a common patterning mechanism in the root and shoot. *Curr. Opin. Plant Biol.* 6, 74–78 (2003).
15. Larkin, J. C., Brown, M. L. & Schiefelbein, J. How do cells know what they want to be when they grow up? Lessons from epidermal patterning in *Arabidopsis. Annu. Rev. Plant Biol.* 54, 403–430 (2003).
16. Koornneef, M., Dellaert, L. W. M. & van der Veen, J. H. EMS- and radiation-induced mutation frequencies at individual loci in *Arabidopsis thaliana. Mutat. Res.* 93, 109–123 (1982).
17. Koornneef, M. The complex syndrome of *ttg* mutants. Arabidopsis *Information Service* 18, 45–51 (1981).
18. Zhang, F., Gonzalez, A., Zhao, M., Payne, C. T. & Lloyd, A. A network of redundant bHLH proteins functions in all TTG1-dependent pathways of *Arabidopsis. Development* 130, 4859–4869 (2003).
19. Schellmann, S. *et al.* TRIPTYCHON and CAPRICE mediate lateral inhibition during trichome and root hair patterning in *Arabidopsis. EMBO J.* 21, 5036–5046 (2002).
20. Wada, T., Tachibana, T., Shimura, Y. & Okada, K. Epidermal cell differentiation in *Arabidopsis* determined by a *myb* homolog, *CPC. Science* 277, 1113–1116 (1997).
21. Kirik, V., Simon, M., Hülskamp, M. & Schiefelbein, J. The *ENHANCER OF TRY AND CPC1 (ETC1)* gene acts redundantly with *TRIPTYCHON* and *CAPRICE* in trichome and root hair cell patterning in *Arabidopsis. Dev. Biol.* 268, 506–513 (2004).
22. Payne, C. T., Zhang, F. & Lloyd, A. M. GL3 encodes a bHLH protein that regulates trichome development in *Arabidopsis* through interaction with GL1 and TTG1. *Genetics* 156, 1349–1362 (2000).

23. Oppenheimer, D. G., Herman, P. L., Sivakumaran, S., Esch, J. & Marks, M. D. A *myb* gene required for leaf trichome differentiation in *Arabidopsis* is expressed in stipules. *Cell* 67, 483–493 (1991).
24. Walker, A. R. et al. The *TRANSPARENT TESTA GLABRA1* locus, which regulates trichome differentiation and anthocyanin biosynthesis in *Arabidopsis*, encodes a WD40 repeat protein. *Plant Cell* 11, 1337–1349 (1999).
25. Szymanski, D. B., Lloyd, A. M. & Marks, M. D. Progress in the molecular genetic analysis of trichome intiation and morphogenesis in *Arabidopsis. Trends Plant Sci.* 5, 214–219 (2000).
26. Esch, J. J. *et al.* A contradictory *GLABRA3* allele helps define gene interactions controlling trichome development in *Arabidopsis. Development* 130, 5885–5894 (2003).
27. Wada, T. *et al.* Role of a positive regulator of root hair development, CAPRICE, in *Arabidopsis* root epidermal cell differentiation. *Development* 129, 5409–5419 (2002).
28. Lee, M. M. & Schiefelbein, J. cell patterning in the *Arabidopsis* root epidermis determined by lateral inhibition with feedback. *Plant Cell* 14, 611–618 (2002).
29. Rerie, W. G., Feldmann, K. A. & Marks, M. D. The *glabra 2* gene encodes a homeo domain protein required for normal trichome development in *Arabidopsis. Genes Dev.* 8, 1388–1399 (1994).
30. Cristina, M. D. *et al.* The *Arabidopsis* Athb-10 (GLABRA2) is an HD-Zip protein required for regulation of root hair development. *Plant J.* 10, 393–402 (1996).
31. Szymanski, D. B., Jilk, R. A., Pollock, S. M. & Marks, M. D. Control of GL2 expression in *Arabidopsis* leaves and trichomes. *Development* 125, 1161–1171 (1998).
32. Ohashi, Y., Ruberti, I., Morelli, G. & Aoyama, T. Entopically additive expression of GLABRA2 alters the frequency and spacing of trichome initiation. *Plant J.* 21, 5036–5046 (2002).
33. Ohashi, Y. *et al.* Modulation of phospholipid signaling by *GLABRA2* in root-hair pattern formation. *Science* 300, 1427–1430 (2003).
34. Walker, J. D., Oppenheimer, D. G., Concienne, J. & Larkin, J. C. *SIAMESE*, a gene controlling the endoreduplication cell cycle in *Arabidopsis thaliana* trichomes. *Development* 127, 3931–3940 (2000).
36. Schnittger, A. *et al.* Ectopic D-type cyclin expression induced not only DNA replication but also cell division in *Arabidopsis* trichomes. *Proc. Natl Acad. Sci. USA* 99, 6410–5415 (2002).
37. Sugimoto-Shirasu, K., Stacey, N. J., Corsar, J., Roberts, K. & McCann, M. C. DNA topoisomerase VI is essential for endoreduplication in *Arabidopsis. Curr. Biol.* 12, 1782–1786 (2002).
38. Jacobsen, S. E., Binkowski, K. A. & Olszewski, N. E. SPINDLY, a tetratricopeptide repeat protein involved in gibberellin signal transduction in *Arabidopsis. Proc. Natl Acad. Sci. USA* 93, 9292–9296 (1996).

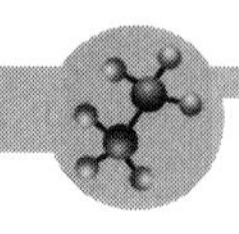

39. Chien, J. C. & Sussex, I. M. Differential regulation of trichome formation on the adaxial and abaxial leaf surfaces by gibberellins and photoperiod in *Arabidopsis thaliana* (L.) Heynh. *Plant Physiol.* 111, 1321–1328 (1996).
40. Telfer, A., Bollman, K. M. & Poethig, R. S. Phase change and the regulation of trichome distribution in *Arabidopsis thaliana. Development* 124, 645–654 (1997).
41. Downes, B. P., Stupar, R. M., Gingerich, D. J. & Vierstra, R. D. The HECT ubiquitin-protein ligase (UPL) family in *Arabidopsis*: UPL3 has a specific role in trichome development. *Plant J.* 35, 729–742 (2003).
42. El Refy, A. *et al.* The *Arabidopsis KAKTUS* gene encodes a HECT protein and controls the number of endoreduplication cycles. *Mol. Genet. Genomics* 270, 403–414 (2004).
43. Hülskamp, M. How plants split hairs. *Curr. Biol.* 10, R308–R310 (2000).
44. Krishnakumar, S. & Oppenheimer, D. G. Extragenic suppressors of the *Arabidopsis zwi-3* mutation identify new genes that function in trichome branch formation and pollen tube growth. *Development* 126, 3079–3088 (1999).
45. Oppenheimer, D. Genetics of plant cell shape. *Curr. Opin. Plant Biol.* 1, 520–524 (1998).
46. Luo, D. & Oppenheimer, D. G. Genetic control of trichome branch number in *Arabidopsis*: the roles of the *FURCA* loci. *Development* 126, 5547–5557 (1999).
47. Perazza, D. *et al.* Trichome cell growth in *Arabidopsis thaliana* can be depressed by mutations in at least five genes. *Genetics* 152, 461–476 (1999).
48. Kirik, V. *et al.* CPR5 is involved in cell proliferation and cell death control and encodes a novel transmembrane protein. *Curr. Biol.* 11, 1891–1895 (2001).
49. Mathur, J. & Chua, N.-H. Microtubule stabilation leads to growth reorientation in *Arabidopsis thaliana* trichomes. *Plant Cell 12*, 465–477 (2000).
50. Kirik, V. et al. The *Arabidopsis TUBULIN-FOLDING COFACTOR A* gene is involved in the control of the α/β-tubulin monomer balance. *Plant Cell* 14, 2265–2276 (2002).
51. Kirik, V. et al. Functional analysis of the tubulin-folding cofactor C in *Arabidopsis thaliana. Curr. Biol.* 12, 1519–1523 (2002).
52. Steinborn, K. *et al.* The *Arabidopsis PILZ* group genes encode tubulin-folding cofactor orthologs required for cell division but not cell growth. *Genes Dev.* 16, 959–971 (2002).
53. Bichet, A., Desnos, T., Turner, S., Grandjean, O. & Höfte, H. BOTERO1 is required for normal orientation of cortical microtubules and anisotropic cell expansion in *Arabidopsis. Plant J.* 25, 137–148 (2001).
54. Webb, M., Jouannic, S., Foreman, J., Linstead, P. & Dolan, L. Cell specification in the *Arabidopsis* root epidermis requires the activity of ECTOPIC ROOT HAIR 3—a katanin-p60 protein. *Development* 129, 123–131 (2002).

55. Burk, D. H., Liu, B., Zhong, R., Morrison, W. H. & Ye, Z. H. A katanin-like protein regulates normal cell wall biosynthesis and cell elongation. *Plant Cell* 13, 807–827 (2001).
56. Torres-Ruiz, R. A. & Jürgens, G. Mutations in the *FASS* gene uncouple pattern formation and morphogenesis in *Arabidopsis* development. *Development* 120, 2967–2978 (1994).
57. Traas, J. *et al.* Normal differentiation patterns in plants lacking microtubular preprophase bands. *Nature* 375, 676–677 (1995).
58. Camilleri, C. *et al.* The *Arabidopsis TONNEAU2* gene encodes a putative novel protein phosphatase 2A regulatory subunit essential for the control of the cortical cytoskeleton. *Plant Cell* 14, 833–845 (2002).
59. Qiu, J. L., Jilk, R., Marks, M. D. & Szymanski, D. B. The *Arabidopsis SPIKE1* gene is required for normal cell shape control and tissue development. *Plant Cell* 14, 101–118 (2002).
60. Brugnera, E. *et al.* Unconventional Rac-GEF activity is mediated through the Dock180–ELMO complex. *Nature Cell Biol.* 4, 574–582 (2002).
61. Deeks, M. J. & Hussey, P. J. Arp2/3 and 'the shape of things to come'. *Curr. Opin. Plant Biol.* 6, 561–567 (2003).
62. Reddy, A. S., Narasimhulu, S. B., Safadi, F. & Golovkin, M. A plant kinesin heavy chain-like protein is a calmodulin-binding protein. *Plant J.* 10, 9–21 (1996).
63. Reddy, A. S. N., Safadi, F., Narasimhulu, S. B., Golovkin, M. & Hu, X. A novel plant calmodulin-binding protein with a kinesin heavy chain motor domain. *J. Biol. Chem.* 271, 7052–7060 (1996).
64. Reddy, A. S. N., Narasimhulu, S. B. & Day, I. S. Structural organization of a gene encoding a novel calmodulin-binding kinesin-like protein from *Arabidopsis. Gene* 204, 195–200 (1997).
65. Song, H., Golovkin, M., Reddy, A. S. & Endow, S. A. *In vitro* motility of AtKCBP, a calmodulin-binding kinesin protein of *Arabidopsis. Proc. Natl Acad. Sci. USA* 94, 322–237 (1997).
66. Deavours, B. E., Reddy, A. S. & Walker, R. A. Ca^{2+}/calmodulin regulation of the *Arabidopsis* kinesin-like calmodulin-binding protein. *Cell Motil. Cytoskeleton* 40, 408–416 (1998).
67. Oppenheimer, D. G. *et al.* Essential role of a kinesin-like protein in *Arabidopsis* trichome morphogenesis. *Proc. Natl Acad. Sci. USA* 94, 6261–6266 (1997).
68. Reddy, V. S., Day, I., Thomas, T. & Reddy, A. S. N. KIC, a novel Ca^{2+} binding protein with one EF-hand motif, interacts with a microtubule motor protein and regulates trichome morphogenesis. *Plant Cell* 16, 185–200 (2004).
69. Folkers, U. *et al.* The cell morphogenesis gene *ANGUSTIFOLIA* encodes a CtBP/BARS-like protein and is involved in the control of the microtubule cytoskeleton. *EMBO J.* 21, 1280–1288 (2002).

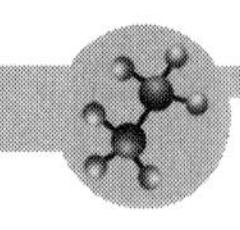

70. Kim, G. T. *et al.* The *ANGUSTIFOLIA* gene of *Arabidopsis*, a plant CtBP gene, regulates leaf-cell expansion, the arrangement of cortical microtubules in leaf cells and expression of a gene involved in cell-wall formation. *EMBO J.* 26, 1267–1279 (2002).
71. Nibu, Y., Zhang, H. & Levine, M. Interaction of a short-range repressors with *Drosophila* CtBP in the embryo. *Science* 280, 101–104 (1998).
72. Matteis, M. D. *et al.* Stimulation of endogenous ADP-ribosylation by brefeldin A. *Proc. Natl Acad. Sci. USA* 91, 1114–1118 (1994).
73. Lippincott-Schwartz, J., Yuan, L. C., Bonifacino, J. S. & Klausner, R. D. Rapid redistribution of Golgi proteins into the ER in cells treated with brefeldin A: evidence for membrane cycling from Golgi to ER. *Cell* 56, 801–813 (1989).
74. Ilgenfritz, H. *et al.* The *Arabidopsis STICHEL* gene is a regulator of trichome branch number and encodes a novel protein. *Plant Physiol.* 131, 643–655 (2003).
75. Szymanski, D. B., Marks, M. D. & Wick, S. M. Organized F-actin is essential for normal trichome morphogenesis in *Arabidopsis. Plant Cell* 11, 2331–2348 (1999).
76. Schwab, B. *et al.* Regulation of cell expansion by the *DISTORTED* genes in *Arabidopsis thaliana*: actin controls the spatial organization of microtubules. *Mol. Genet. Genomics* 269, 350–360 (2003).
77. Mathur, J., Spielhofer, P., Kost, B. & Chua, N.-H. The actin cytoskeleton is required to elaborate and maintain spatial patterning during trichome cell morphogenesis in *Arabidopsis thaliana. Development* 126, 5559–5568 (1999).
78. Mathur, J. *et al. Arabidopsis CROOKED* encodes for the smallest subunit of the ARP2/3 complex and controls cell shape by region specific fine F-actin formation. *Development* 130, 3137–3146 (2003).
79. Mathur, J., Mathur, N., Kernebeck, B. & Hülskamp, M. Mutations in actin-related proteins 2 and 3 affect cell shape development in *Arabidopsis. Plant Cell* 15, 1632–1645 (2003).
80. Le, J., El-Assal Sel, D., Basu, D., Saad, M. E. & Szymanski, D. B. Requirements for *Arabidopsis* ATARP2 and ATARP3 during epidermal development. *Curr. Biol.* 13, 1341–1347 (2003).
81. Li, S., Blanchoin, L., Yang, Z. & Lord, E. M. The putative *Arabidopsis* arp2/3 complex controls leaf cell morphogenesis. *Plant Physiol.* 132, 2034–2044 (2003).
82. Mullins, R. D., Heuser, J. A. & Pollard, T. D. The interaction of Arp2/3 complex with actin: nucleation, high affinity pointed end capping, and formation of branching networks of filaments. *Proc. Natl Acad. Sci. USA* 95, 6181–6186 (1998).
83. Svitkina, T. M. & Borisy, G. G. ARP2/3 complex and actin depolymerizing factor/cofilin in dendritic organization and treadmilling of actin filament array in lamellipodia. *J. Cell Biol.* 145, 1009–1026 (1999).
84. Mathur, J. & Hülskamp, M. Signal transduction: Rho-like proteins in plants. *Curr. Biol.* 12, R526–R528 (2002).

85. Smith, L. G. Cytoskeletal control of plant cell shape: getting the fine points. *Curr. Opin. Plant Biol.* 6, 63–73 (2003).
86. Yang, Z. Small GTPases: versatile signaling switches in plants. *Plant Cell* 14 (Suppl.) 375–388 (2002).
87. De Veylder, L. *et al.* Functional analysis of cyclin-dependent kinase inhibitors of *Arabidopsis. Plant Cell* 13, 1653–1668 (2001).
88. Hiromura, K., Pippin, J. W., Fero, M. L., Roberts, J. M. & Shankland, S. J. Modulation of apoptosis by the cyclin-dependent kinase inhibitor p27(Kip1). *J. Clin. Invest.* 103, 597–604 (1999).
89. Wang, H. *et al.* ICK1, a cyclin-dependent protein kinase inhibitor from *Arabidopsis thaliana* interacts with both Cdc2a and CycD3, and its expression is induced by abscisic acid. *Plant J.* 15, 501–510 (1998).
90. Jasinski, S. *et al.* The CDK inhibitor NtKIS1a is involved in plant development, endoreduplication and restores normal development of cyclin D3;1-overexpressing plants. *J. Cell Sci.* 115, 973–982 (2002).
91. Schnittger, A., Weinl, C., Bouyer, D., Schobinger, U. & Hülskamp, M. Misexpression of the cyclin-dependent kinase inhibitor ICK1/KRP1 in single-celled *Arabidopsis* trichomes reduces endoreduplication and cell size and induces cell death. *Plant Cell* 15, 303–315 (2003).
92. Glazebrook, J. Genes controlling expression of defense responses in *Arabidopsis*—2001 status. *Curr. Opin. Plant Biol.* 4, 301–308 (2001).
93. Haughn, G. W. & Somerville, C. R. Genetic control of morphogenesis in *Arabidopsis. Dev. Genet.* 9, 73–89 (1988).
94. Potikha, T. & Delmer, D. A mutant of *Arabidopsis thaliana* displaying altered patterns of cellulose deposition. *Plant J.* 7, 453–460 (1995).

As mentioned previously, cell division is an important process in differentiation and development. The cell division cycle is composed of multiple stages:

- *G1—Growth phase 1, in which proteins are made for cell division*

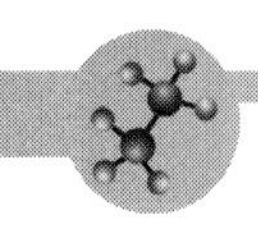

- *S—Synthesis, in which the cell's entire complement of DNA is replicated*
- *G2—Growth phase 2, in which more proteins are made for cell division*
- *M—Mitosis, in which, through several phases, DNA condenses into chromosomes that are paired, separated, and subsequently distributed to each daughter cell. The cytoplasm is also distributed, in a process known as cytokinesis, usually evenly to each daughter cell.*

There are some differences between plants and animals in the cell division process. Animal cells have a pair of organelles called centrioles, which are used to help pull the chromosomes apart, while plants lack these structures. Animal cells also form a groove in the cell called a cleavage furrow during cytokinesis, while plant cells form and unite a series of vesicles to form a cell plate and new cell wall between daughter cells. Animal cells generally have only two sets of chromosomes that can be replicated, while plant cells can have many more. In the following article, Keith Lindsey and Jennifer Topping discuss how the process of cell division in plants influences all aspects of cell differentiation and development. —CCF

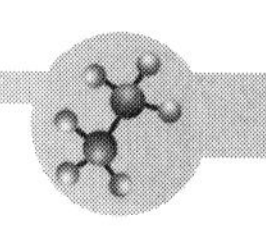

"On the Relationship Between the Plant Cell and the Plant"

by Keith Lindsey and Jennifer Topping

***Seminars in Cell & Developmental Biology*, 1998**

A fundamental difference between the early developmental processes of plants and animals is the manner in which cellular pattern becomes organized. In mammals, for example, a major feature of embryo development is the migration of cells to generate three-dimensional shape. The division of the cells leads to cell separation, though various types of intercellular communication become established in different cellular contexts.[1] In flowering plants, however, the only cell migration to occur is during fertilization, when the pollen grain is released from the anther and the pollen tube and its cellular components grow down the style towards the embryo sac. In other phases of the life-cycle, independent cell migration does not occur. While many of the essential components of cell cycle control are functionally conserved between plants and animals, cytokinesis is quite different at the structural level and presumably, therefore, also at the molecular level.

Primitive filamentous algae such as *Klebsormidium* divide by fission, although *Spirogyra*, which shows division predominantly by this mechanism,[2] also exhibits elements of cell plate formation,[3] although this may be functionally distinct from that of higher plants.[4] In any case, the vast majority of plant species generate a new wall from the inside out (centrifugally).[5] The thalloid

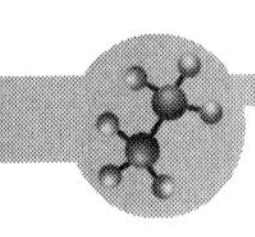

freshwater green alga *Coleochaete*, however, is unusual in clearly exhibiting elements both of fission and of cell plate formation: anticlinal and periclinal divisions are carried out by different mechanisms. For periclinal divisions, the new wall is formed centripetally, i.e. from the outside inwards. This is achieved by the movement of vesicles away from the old wall and is considered to represent a form of furrowing reminiscent of the fission of *Spirogyra*. For the anticlinal divisions, producing a new radial wall, wall material is deposited centrifugally, as during phragmoplast formation in higher plants.[6] We may conclude that the mechanisms of cytokinesis in modern higher plant cells are relatively recent in evolutionary terms, but originated in ancestors common to both the higher plants and the freshwater green algae such as *Coleochaete*.

Presumably there has been a strong selection pressure acting on the centrifugal mode of cytokinesis since it has been adopted so widely. We can ask, what advantages have been conferred on plants that generate a phragmoplast, and in particular, what developmental consequences have ensued? And what does this tell us about the relationship between cell and organism in higher plants, with respect to developmental processes? The position of the new cell wall, its composition, and the extent of the continuity between the new cells are each variables that may influence the expression of developmental pathways.

Cell Division Planes Can Influence Cell Fate

While the products of a mitotic division are typically genetically identical, their fates may be quite different.

Commonly, this differential fate correlates with an asymmetric separation of the mother cell cytoplasm. For example, the zygote of the brown alga *Fucus* divides to produce progenitor cells of the thallus and rhizoid, respectively. The asymmetry of cell fate is a response to the asymmetric distribution of an environmental signal, light, leading to a polar secretion of cell wall components,[7] which themselves play a determinative role in the regulation of cell fate.[8]

In *Arabidopsis*, the immediate division products of the zygote form either the embryo proper or the suspensor, and differences in gene expression in the two cell population are clear. For example, we have identified, by promoter trapping, genes that are expressed in the *Arabidopsis* embryo but not in the suspensor[9] and the *HYDRA1* gene product is required for the establishment of normal cell shape in the embryo proper but not the suspensor.[10] A homeobox gene *ATLM1* has been identified that is expressed in the apical but not basal cell of the embryo.[11] The proteoglycan cell wall components (arabinogalactan-proteins, AGPs) of the *Brassica napus* embryo proper and suspensor are also distinct, as revealed by the suspensor-specific binding of the JIM8 antibody to the suspensor.[12] The possibility exists that pectinaceous or other carbohydrate components of the plant cell wall may play critical roles in the regulation of differentiation and development, not only in embryos but also in roots through effects on cell expansion,[13, 14] and in floral organs.[15]

The products of microspore mitosis become the vegetative and generative cells of the pollen grain with distinct functions.[16] The asymmetric division of the

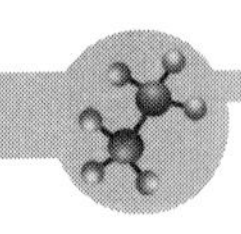

microspore appears to play a determinative role in the expression of daughter cell-specific gene expression patterns, since the manipulation of division symmetry by colchicine treatment directly influences vegetative cell-specific gene expression.[17]

In the *Arabidopsis* root, products of an asymmetric division of an initial cell differentiate into the cortical and endodermal layers, respectively, and the *scarecrow* mutant, in which this division is absent, produces only a single cell layer instead of the usual two.[18] Interestingly, the single new layer displays a combination of characteristics each derived from either the endodermis or cortex, indicating that a positional signalling system determines the identity of radial cell layers.

Therefore, at the level of cell differentiation, asymmetry of cell division may provide either a pattern of distribution of cytoplasmic factors, or a patterning of cells in different positions, leading to the asymmetry of gene expression for the activation of alternative developmental pathways. This is clearly seen in the *Fucus* system, in which progress is advancing rapidly to identify the wall components that influence cell fate so dramatically. There is now also good evidence of differential plasmodesmatal connectivity between plant cells in different tissues, opening the possibility of a role for dynamic intercellular communication (differential gating between cells) in the regulation of differentiation.[19, 20] Interestingly, the more primitive filamentous algae such as *Klebsormidium*, that divide by fission, appear not to have plasmodesmata and exhibit no cell differentiation, except in the formation of asexual flagellate zoospores.[21]

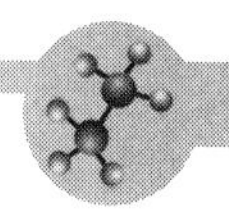

Therefore, the plant cell plate can carry information as structural components, and may regulate the translocation of information between cells via the integral plasmodesmata, to mediate cell differentiation pathways. One can therefore envisage a selective advantage conferred on organisms with cell plates as a means of generating polarity or positional information, acting as a local regulator of the fates of adjacent cells.

Relationship Between Cell Division and Morphogenesis

Meristems are typically highly ordered structures, suggesting an important role for regulated cell division patterns in morphogenesis. In *Arabidopsis*, for example, the patterning of cells in the embryo and root is highly stereotyped and predictable. Even in the shoot apical meristem (SAM), which is less stereotyped, the regulation of division in the L1 and L2 layers is such that the predominantly anticlinal divisions generate distinct cell layers that rarely invade each other. Within the shoot tip, there are regions with cell groups (such as the central zone and the flanking regions) that are predictably different in their rates of division.[22, 23]

From an evolutionary perspective, it is possible to see a correlation between regulated orientations of cell plate formation and a more complex morphology. While some unicellular organisms such as the siphonous (coenocytic) green alga *Acetabularia* can generate quite complex morphologies in the absence of cell divisions, structural complexity in multicellular organisms is associated with more complex patterns of cell division.

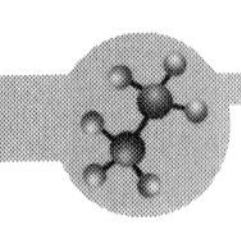

Simple filamentous growth in algae such as the zygnematales is achieved by cell division in a single plane; and in mosses, there is a filamentous protonema that grows by tip-growth. In neither case are pre-prophase bands formed. In switching to a more leafy shoot-like growth habit, however, the moss caulonema does generate pre-prophase bands and associated multiple cell layers.[24] There is therefore a correlation between three-dimensional growth and pre-prophase band formation, as a mechanism that may allow control over the plane of division. In the *tonneau* mutant (allelic to *fass*), there is a failure of pre-prophase band formation[25] and a corresponding defect in the control of cell shape from early in embryogenesis. The result is that morphogenesis is defective, but significantly cell differentiation and correct apical-basal patterning can proceed despite this defect.[26]

The *hydra* mutant[10] also exhibits serious defects in cell shape, with an inability to undergo correct axial (longitudinal) cell expansion, even when grown in the dark. Morphogenesis is abnormal, but like *fass/tonneau*, *hydra* mutants nevertheless can undergo cell differentiation, as revealed by the expression of cell type-specific molecular markers. However, histogenesis is severely affected. Other cell shape-defective mutants such as *emb30*, *keule* and *knolle* have morphological defects. There have also been identified other mutants with defective cellular patterning, in spatially much more restricted positions than is seen in the more pleiotropic mutants described above. These include *monopteros*,[27] *hobbit*,[28] *shortroot*,[29] and *scarecrow*,[18] among others, in which aberrant cellular

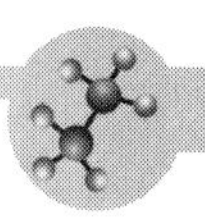

patterning in the embryonic root correlates with defective root morphogenesis.

Together these mutants show that abnormalities in the construction or positioning of new cell walls, either through defective pre-prophase band formation, defective cell expansion or defective secretion of essential components to the cell wall (or a combination of these) can result in abnormal morphogenesis. Defects in the rate of cell division *per se* do not appear to be the fundamental defects in these mutants.

A number of mutants suggest the importance of correct rates of cell division in morphogenesis. The *clavata* mutants exhibit abnormally high rates of cell division at the shoot apex, leading to increased numbers of cells in the shoot apex and to dramatic morphological defects.[30–32] These genes are therefore required to down-regulate the rate of cell division in the SAM. Other loci are required to ensure sufficient cell division for correct organogenesis, including *KNOTTED* of maize,[33, 34] the related *SHOOTMERISTEMLESS* of *Arabidopsis*[35] and *WUSCHEL*, also of *Arabidopsis*.[36] Clearly, cell division activity is under tight genetic control. Disruption of any of these components can result in defective development.

However, there are still interesting questions to be answered concerning the role of the rate of cell division in growth and morphogenesis. It has been possible to manipulate experimentally the rate of cell division in roots yet, surprisingly, leave root morphogenesis unperturbed.[37, 38] The implication of these results is that control over the rate of cell division can be uncoupled from control of correct morphogenesis. The data may be interpreted as demonstrating that plant cells can in some way sense the rate of cell division of their neighbours, and differentiate or divide according to the context in which they are found. Laser ablation of root cells has demonstrated clearly that cells monitor the identity and division activity of their neighbours:[39] they will divide if a neighbour is dead, and differentiate according to their new position following ectopic division. In the light of these observations, it may be that a defect in *clavata* mutants is in the ability of the cells in the shoot apex to monitor their local environment and to differentiate correctly.

Relationship Between Cell Division and Pattern Formation

The predictability of cellular pattern in the *Arabidosis* embryo begs the question of whether there is a causal relationship between correct cell division and pattern formation at the embryo or seedling level, in this species and in others. As alluded to earlier, there is an increasing amount of evidence that apical-basal pattern formation is regulated independently of correct cell morphogenesis and division.

Embryogenesis in maize differs from that of *Arabidopsis* not only in the organization of the mature embryo, but also in the lack of stereotyped cell filing.[40] Clearly, maize embryos are of a predictable final shape, but this is generated in the absence of highly ordered cell files such as are seen in *Arabidopsis*. The *tangled-1* mutant of maize is characterized by severely defective planes of cell division in the leaf, although the shape of

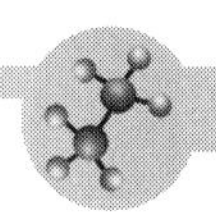

the leaf is unaffected.[41] Embryonic mutants of *Arabidopsis* such as *fass*,[26] *knolle*,[42] *keule*[41] and *hydra*[10] show pleiotropic defects in cellular organization, and yet still are able to establish correct apical-basal polarity: cotyledons, hypocotyls and root are produced in the correct relative positions, albeit of abnormal shapes. Using promoter trapping, we have generated novel molecular markers expressed in specific regions along the apical-basal axis of the *Arabidopsis* seedling[9, 44] and used them to investigate the apical-basal patterning of two mutants, *hydra* and *emb30/gnom*.[45]

The *EXORDIUM (EXO)* marker is expressed in the cotyledons, shoot apex, and primary and lateral root apices of seedlings, and expression in *hydra* seedlings in essentially the same pattern as in the wild type.[45] In *emb30* seedlings with the least severe mutant phenotypes, expression is restricted to the cotyledonary regions and not observed in the position of root and shoot meristems, consistent with the interpretation that these *emb30* mutants have fused cotyledons but no meristems.

A second marker, *COLUMELLA (COL)*, is expressed specifically in the root cap of both primary and lateral roots. It is active in the least abnormal *hydra* root tips, consistent with the anatomical observation that a root cap is present in such roots. *COL* is not expressed in *emb30* seedlings.

A third marker, *POLARIS (PLS)*, is first expressed in the basal half of the embryo from the heart stage onward and in the seedling and mature plant root tip. This marker is also active in the basal region of the

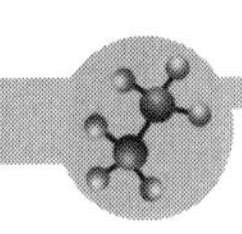

hydra embryos, although *hydra* has no morphologically obvious embryonic root; and in the basal region of the *hydra* seedling, in the absence of correct root organogenesis. This unexpected observation was also made for *emb30* mutants carrying this marker, even though the mutants have no root meristem.[46] The precise level and position of GUS activity showed some variability between *emb30* seedling siblings, with the most aberrant (least polarized) seedlings showing the least polarized expression.

One interpretation of the pattern of *PLS* expression is that it is regulated in a way that is independent of anatomical or structural facets of root cell differentiation *per se*, but rather reflects a biochemical differentiation of those cells in a position-related manner. This suggests that *PLS* may be activated by a signalling pathway that regulates position-dependent gene expression in the embryonic and seedling root, and that, in the severest *emb30* mutant seedlings, this signalling pathway is disrupted. These observations suggest: (1) that *EXO* and *COL* on the one hand and *PLS* on the other resolve different components of the pathway of root development; and (2) that at least some components of apical-basal patterning may be generated in the absence of correct cellular organization.

Conclusions

Higher plants grow predominantly in the longitudinal axis, either upwards towards the light above ground, or down towards water and nutrients beneath ground level. Much of this growth is due to cell expansion. New

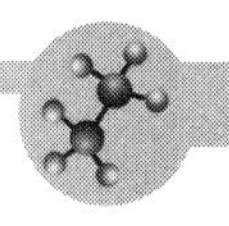

cell plates, generated in the meristems, ensure a homeostasis of cell size, and may form in response to cell expansion[47] and mutants in which cell shape control is defective characteristically are dwarfed. A strong selective advantage for expansion growth that is independent of extended (or correct) cell division presumably resides in the high value for germinating seedlings in reaching the light as quickly as possible; cell division activity for body plan formation has been completed well before seed dormancy. Subsequent vegetative growth from meristems is largely a reiteration of cellular patterning established during embryogenesis. However, the mechanisms that define apical-basal patterning independent of correct cell division or cell shape remain obscure.

If correct cell division patterns are not strictly required for apical-basal patterning, as in *fass* and *hydra*, they do appear to be important in generating a correct radial pattern. This may be because of the timing of the first lesion in these mutants, in the pre-globular embryo when the radial pattern is being established, but it is also possible that cell fate determining signals are more readily disrupted in the radial than the apical-basal plane in these mutants. The selective advantage of correct patterns of cell division might therefore primarily be in the establishment of correct histogenesis and tissue function. Furthermore, newly emerging evidence implicates the cell plate in regulating cell fate, through differential cell wall composition or differential gating between cells via the plasmodesmata. Plant cells probably continuously monitor their

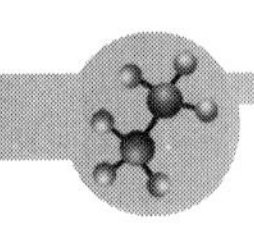

environment, and regulate their rate of division and fate accordingly.[18, 39] It may be that in some mutants, such as *fass* or *hydra*, cells can monitor their position in the apical-basal axis but less readily in the radial axis. Severe alleles of *emb30*, on the other hand, may be defective in apical-basal signalling, as evidenced by a disrupted expression pattern of polarity markers such as *POLARIS*.

Plant cells are not independent structures. It is not possible to culture single cells for very long (they rapidly form multicellular aggregates), and their fate depends very much on the context in which they are found. It is becoming clearer how the role of cell division also differs according to developmental context, and different developmental processes appear to be independently regulated, relying on mechanisms that appear to have their origins in aquatic algae. New mutants and molecular markers, such as those generated by promoter trapping, should help us further in unravelling the components of these mechanisms.

References

1. Greenwald I, Rubin GM (1992) Making a difference: the role of cell-cell interactions in establishing separate identities for equivalent cells. Cell 68:271–281
2. Goto Y, Ueda K 1988 Microfilament bundles of F-actin in *Spirogyra* observed by fluorescence microscopy. Planta 173:442–446
3. McIntosh K, Pickett-Heaps JD, Gunning BES (1995) Cytokinesis in *Spirogyra*: integration of cleavage and cell plate formation. Int J Plant Sci 156:1–8
4. Sawitzky H, Grolig F (1995) Phragmoplast of the green alga *Spirogyra* is functionally distinct from the higher plant phragmoplast. J Cell Biol 130:1359–1371
5. Staehelin LA, Hepler PK (1996) Cytokinesis in higher plants. Cell 84:821–824
6. Marchant HJ, Pickett-Heaps JD (1973) Mitosis and cytokinesis in *Coleochaete scutata*. J Phycol 9:461–471

7. Shaw SL, Quatrano RS (1996) The role of targeted secretion in the establishment of cell polarity and the orientation of the division plane in *Fucus* zygotes. Development 122:2623–2630
8. Berger F, Taylor A, Brownlee C (1994) Cell fate determination by the cell wall in early *Fucus* development. Science 263:1421–1423
9. Topping JF, Agyeman F, Henricot B, Lindsey K (1994) Identification of molecular markers of embryogenesis in *Arabidopsis thaliana* by promoter trapping. Plant J 5:895–903
10. Topping JF, May VJ, Muskett PR, Lindsey K (1997) Mutations in the *HYDRA1* gene of *Arabidopsis* perturb cell shape and disrupt embryonic and seedling morphogenesis. Development 124:4415–4424
11. Lu P, Porat R, Nadeau JA, O'Neill SD (1996) Identification of a meristem L1 layer-specific gene in *Arabidopsis* that is expressed during embryonic pattern formation and defines a new class of homeobox genes. Plant Cell 8:2155–2168
12. Pennell RI, Janniche L, Kjellbom P, Scofield GN, Peart JM, Roberts K (1991) Developmental regulation of a plasma membrane arabinogalactan protein in oilseed rape flowers. Plant Cell 3:1317–1326
13. Fry SC, Aldington S, Hetherington PR, Aitken J (1993) Oligosaccharides as signals and substrates in the plant cell wall. Plant Physiol 103:1–5
14. Willats WGT, Knox JP (1996) A role for arabinogalactan-protein in plant cell expansion: evidence from studies on the interaction of β-glucosyl Yariv reagent with seedlings of *Arabidopsis thaliana.* Plant J 9:919–925
15. Kawaguchi K, Shibuya N, Ishii T (1996) A novel tetrasaccharide, with a structure similar to the terminal sequence of an arabinogalactan-protein, accumulates in rice anthers in a stagespecific manner. Plant J 9:777–785
16. Terasaka O, Niitsu T (1987) Unequal cell division and chromatin differentiation in pollen grain cells. I. Centrifugal, cold and caffeine treatments. Bot Mag Tokyo 100:205–216
17. Eady C, Lindsey K, Twell D (1995) The significance of microspore division asymmetry for vegetative cell-specific transcription and generative cell differentiation. Plant Cell 7:65–74
18. Di Laurenzio L, Wysocka-Diller J, Malamy J, Pysh L, Helariutta Y, Freshour G, Hahn MG, Feldmann KA, Benfey PN (1996) The *SCARECROW* gene regulates an asymmetric cell division that is essential for generating the radial organization of the Arabidopsis root. Cell 86:423–433
19. Duckett CM, Oparka KJ, Prior DAM, Dolan L, Roberts K (1994) Dye-coupling in the root epidermis of *Arabidopsis* is progressively reduced during development. Development 120:3247–3255
20. Lucas WJ, Bouché-Pillon S, Jackson PJ, Nguyen L, Baker L, Ding B, Hake S (1995) Selective trafficking of KNOTTED1 homeodomain protein and its mRNA through plasmodesmata. Science 270:1980–1983

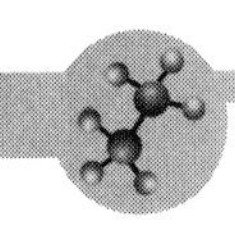

21. Mattox KR, Stewart KD (1984) Classification of the green algae: a concept based on comparative cytology, in Systematics of the Green Algae, (Irvine J, ed.) Academic Press: London
22. Steeves TA, Sussex IM (1989) Patterns in Plant Development. Cambridge University Press: Cambridge
23. Meyerowitz EM (1997) Genetic control of cell division patterns in developing plants. Cell 88:299–308
24. Doonan JH (1991) The cytoskeleton and moss morphogenesis, in The Cytoskeletal Basis of Plant Growth and Form (Lloyd C, ed.) pp. 289–301. Academic Press: London
25. Traas J, Bellini C, Nacry P, Kronenberger J, Bouchéz D, Caboche M (1995) Normal differentiation patterns in plants lacking microtubular preprophase bands. Nature 375:676–677
26. Torres-Ruiz RA, Jürgens G (1994) Mutations in the *FASS* gene uncouple pattern formation and morphogenesis in *Arabidopsis* development. Development 120:2967–2978
27. Berleth T, Jürgens G (1993) The role of the *monopteros* gene in organising the basal body region of the *Arabidopsis* embryo. Development 118:575–587
28. Willemsen V, Wolkenfelt H, de Vrieze G, Weisbeek P, Scheres B (1998) The *HOBBIT* gene is required for formation of the root meristem in the *Arabidopsis* embryo. Development 125:521–531
29. Scheres B, Di Laurenzio L, Willemsen V, Hauser M-T, Jan- maat K, Weisbeek P, Benfey PN (1995) Mutations affecting the radial organisation of the *Arabidopsis* root display specific defects throughout the embryonic axis. Development 121:53–62
30. Leyser HMO, Furner IJ (1992) Characterization of three shoot apical meristem mutants of *Arabidopsis thaliana*. Development 116:397–403
31. Clark SE, Running MP, Meyerowitz EM (1993) *CLAVATA1*, a regulator of meristem and flower development in *Arabidopsis*. Development 119:397–418
32. Clark SE, Running MP, Meyerowitz EM (1995) *CLAVATA3* is a specific regulator of shoot and floral meristem development affecting the same processes as *CLAVATA1*. Development 121:2057–2067
33. Vollbrecht E, Veit B, Sinha N, Hake S (1991) The developmental gene *KNOTTED1* is a member of the maize homeobox gene family. Nature 350:241–243
34. Jackson D, Veit B, Hake S (1994) Expression of maize *KNOTTED1* related homeobox genes in the shoot apical meristem predicts patterns of morphogenesis in the vegetative shoot. Development 120:405–413
35. Long JA, Moan EI, Medford JI, Barton MK (1996) A member of the KNOTTED class of homeodomain proteins encoded by the STM gene of Arabidopsis. Nature 379:66–69

36. Laux T, Mayer KFX, Berger J, Jürgens G (1996) The *WUSCHEL* gene is required for shoot and floral meristem integrity in *Arabidopsis*. Development 122:87–96
37. Hemerley A, de Almeida Engler J, Bergounioux C, Van Montagu M, Engler G, Inzé D, Ferreira P (1995) Dominant negative mutants of the Cdc2 kinase uncouple cell division from iterative plant development. EMBO J 14:3925–3936
38. Doerner P, Jorgensen J-E, You R, Steppuhn J, Lamb C (1996) Control of root growth and development by cyclin expression. Nature 380:520–523
39. van den Berg C, Willemsen V, Hage W, Weisbeek P, Scheres B (1995) Cell fate in the *Arabidopsis* root meristem determined by directional signalling. Nature 378:62–65
40. Clark JK, Sheridan WF (1991) Isolation and characterization of 51 *embryo-specific* mutations of maize. Plant Cell 3:935–951
41. Smith LG, Hake S, Sylvester AW (1996) The *tangled-1* mutation alters cell division orientations throughout maize leaf development without altering leaf shape. Development 122:481–489
42. Lukowitz W, Mayer U, Jürgens G (1996) Cytokinesis in the Arabidopsis embryo involves the syntaxin-related KNOLLE gene product. Cell 84:61–71
44. Lindsey K, Wei, W, Clarke, MC, McArdle, HF, Rooke, LM Topping, JF (1993) Tagging genomic sequences that direct transgene expression by activation of a promoter trap in plants. Transgenic Res 2:33–47
45. Topping JF, Lindsey K (1997) Promoter trap markers differentiate structural and positional components of polar development in *Arabidopsis*. Plant Cell 9:1713–1725
46. Mayer U, Büttner G, Jurgens G (1993) Apical-basal pattern formation in the *Arabidopsis* embryo: studies on the role of the *gnom* gene. Development 117:149–162
47. Jacobs T (1997) Why do plant cells divide? Plant Cell 9:1021–1029

All of the information needed to make an organism is contained in the genome, the total of all of a cell's DNA. When active, genes are transcribed into messenger RNA, which is then translated into proteins, which are the expressed forms of

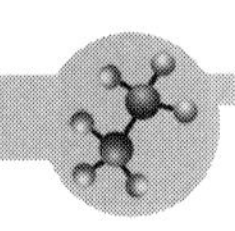

genes that carry out a variety of functions. However, not all genes are active at once. Indeed, the coordinated activity of various genes is essential in normal development as well as everyday functions. So, how do cells regulate this coordinated genetic activity?

For example, in bacteria, the lactose (lac) operon system is a system of genes whose activity is coordinated to make the enzyme that metabolizes the sugar lactose only when lactose is present. In the lac operon, several genes are found in tandem along a section of DNA. The regions of DNA include a regulatory gene (code for a repressor protein), the promoter region (where the enzyme RNA polymerase binds to make messenger RNA), the operator region (where the repressor protein binds), and the structural genes (code for the enzyme that breaks down lactose). The regulatory gene makes the repressor protein. If the enzyme is not needed, then the repressor binds to the operator region and prevents transcription of the structural gene by RNA polymerase. If the enzyme is needed, a substance binds to the repressor protein and prevents it from binding to the operator region, thereby allowing the structural gene to be transcribed.

The lac operon is only one type of known genetic control system; there may be others. If operon systems predominate the genome, then there are many genes that are used primarily as switches for the expression of other genes. In the following article, Nelson Lau and David Bartel discuss how a new type of gene regulation by RNA occurs, which is important both in development and disease treatment. —CCF

"Censors of the Genome"
by Nelson C. Lau and David P. Bartel
***Scientific American*, August 2003**

Biologists have been surprised to discover that most animal and plant cells contain a built-in system to silence individual genes by shredding the RNA they produce. Biotech companies are already working to exploit it.

Observed on a microscope slide, a living cell appears serene. But underneath its tranquil facade, it buzzes with biochemical chatter. The DNA genome inside every cell of a plant or animal contains many thousands of genes. Left to its own devices, the transcription machinery of the cell would express every gene in the genome at once: unwinding the DNA double helix, transcribing each gene into single-stranded messenger RNA and, finally, translating the RNA messages into their protein forms.

No cell could function amid the resulting cacophony. So cells muzzle most genes, allowing an appropriate subset to be heard. In most cases, a gene's DNA code is transcribed into messenger RNA only if a particular protein assemblage has docked onto a special regulatory region in the gene.

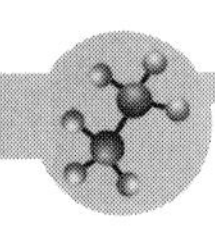

Some genes, however, are so subversive that they should never be given freedom of expression. If the genes from mobile genetic elements were to successfully broadcast their RNA messages, they could jump from spot to spot on the DNA, causing cancer or other diseases. Similarly, viruses, if allowed to express their messages unchecked, will hijack the cell's protein production facilities to crank out viral proteins.

Cells have ways of fighting back. For example, biologists long ago identified a system, the interferon response, that human cells deploy when viral genes enter a cell. This response can shut off almost all gene expression, analogous to stopping the presses. And just within the past several years, scientists have discovered a more precise and—for the purposes of research and medicine—more powerful security apparatus built into nearly all plant and animal cells. Called RNA interference, or RNAi, this system acts like a censor. When a threatening gene is expressed, the RNAi machinery silences it by intercepting and destroying only the offender's messenger RNA, without disturbing the messages of other genes.

As biologists probe the modus operandi of this cellular censor and the stimuli that spur it into action, their fascination and excitement are growing. In principle, scientists might be able to invent ways to direct RNA interference to stifle genes involved in cancer, viral infection or other diseases. If so, the technology could form the basis for a new class of medicines.

Meanwhile researchers working with plants, worms, flies and other experimental organisms have already learned how to co-opt RNAi to suppress nearly

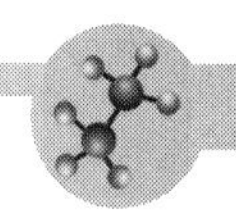

any gene they want to study, allowing them to begin to deduce the gene's purpose. As a research tool, RNAi has been an immediate success, allowing hundreds of laboratories to tackle questions that were far beyond their reach just a few years ago.

Whereas most research groups are using RNA interference as a means to an end, some are investigating exactly how the phenomenon works. Other labs (including our own) are uncovering roles for the RNAi machinery in the normal growth and development of plants, fungi and animals—humans among them.

A Strange Silence

The first hints of the RNAi phenomenon surfaced 13 years ago. Richard A. Jorgensen, now at the University of Arizona, and, independently, Joseph Mol of the Free University of Amsterdam inserted into purple-flowered petunias additional copies of their native pigment gene. They were expecting the engineered plants to grow flowers that were even more vibrantly violet. But instead they obtained blooms having patches of white.

Jorgensen and Mol concluded that the extra copies were somehow triggering censorship of the purple pigment genes—including those natural to the petunias—resulting in variegated or even albino-like flowers. This dual censorship of an inserted gene and its native counterpart, called co-suppression, was later seen in fungi, fruit flies and other organisms.

Clues to the mystery of how genes were being silenced came a few years later from William G. Dougherty's lab at Oregon State University. Dougherty

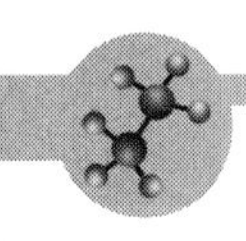

and his colleagues started with tobacco plants that had been engineered to contain within their DNA several copies of the CP (coat protein) gene from tobacco etch virus. When these plants were exposed to the virus, some of the plants proved immune to infection. Dougherty proposed that this immunity arose through co-suppression. The plants apparently reacted to the initial expression of their foreign CP genes by shutting down this expression and subsequently also blocking expression of the CP gene of the invading virus (which needed the coat protein to produce an infection). Dougherty's lab went on to show that the immunity did not require synthesis of the coat protein by the plants; something about the RNA transcribed from the CP gene accounted for the plants' resistance to infection.

The group also showed that not only could plants shut off specific genes in viruses, viruses could trigger the silencing of selected genes. Some of Dougherty's plants did not suppress their CP genes on their own and became infected by the virus, which replicated happily in the plant cells. When the researchers later measured the RNA being produced from the CP genes of the affected plants, they saw that these messages had nearly vanished—infection had led to the CP genes' inactivation.

Meanwhile biologists experimenting with the nematode *Caenorhabditis elegans*, a tiny, transparent worm, were puzzling over their attempts to use "antisense" RNA to inactivate the genes they were studying. Antisense RNA is designed to pair up with

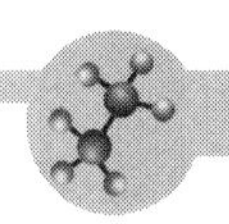

a particular messenger RNA sequence in the same way that two complementary strands of DNA mesh to form a double helix. Each strand in DNA or RNA is a chain of nucleotides, genetic building blocks represented by the letters A, C, G and either U (in RNA) or T (in DNA). C nucleotides link up with Gs, and As pair with Us or Ts. A strand of antisense RNA binds to a complementary messenger RNA strand to form a doublestranded structure that cannot be translated into a useful protein.

Over the years, antisense experiments in various organisms have had only spotty success. In worms, injecting antisense RNAs seemed to work. To everyone's bewilderment, however, "sense" RNA also blocked gene expression. Sense RNA has the same sequence as the target messenger RNA and is therefore unable to lock up the messenger RNA within a double helix.

The stage was now set for the eureka experiment, performed five years ago in the labs of Andrew Z. Fire of the Carnegie Institution of Washington and Craig C. Mello of the University of Massachusetts Medical School. Fire and Mello guessed that the previous preparations of antisense and sense RNAs that were being injected into worms were not totally pure. Both mixtures probably contained trace amounts of double-stranded RNA. They suspected that the double-stranded RNA was alerting the censors.

To test their idea, Fire, Mello and their colleagues inoculated nematodes with either single- or double-stranded RNAs that corresponded to the gene *unc-22*,

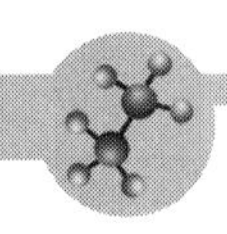

which is important for muscle function. Relatively large amounts of single-stranded *unc-22* RNA, whether sense or antisense, had little effect on the nematodes. But surprisingly few molecules of double-stranded *unc-22* RNA caused the worms—and even the worms' offspring—to twitch uncontrollably, an unmistakable sign that something had started interfering with *unc-22* gene expression. Fire and Mello observed the same amazingly potent silencing effect on nearly every gene they targeted, from muscle genes to fertility and viability genes. They dubbed the phenomenon "RNA interference" to convey the key role of double-stranded RNA in initiating censorship of the corresponding gene.

Investigators studying plants and fungi were also closing in on double-stranded RNA as the trigger for silencing. They showed that RNA strands that could fold back on themselves to form long stretches of double-stranded RNA were potent inducers of silencing. And other analyses revealed that a gene that enables cells to convert single-stranded RNA into double-stranded RNA was needed for co-suppression. These findings suggested that Jorgensen and Mol's petunias recognized the extra pigment genes as unusual (through a mechanism that is still mysterious) and converted their messenger RNAs into double-stranded RNA, which then triggered the silencing of both the extra and native genes. The concept of a double-stranded RNA trigger also explains why viral infection muzzled the CP genes in

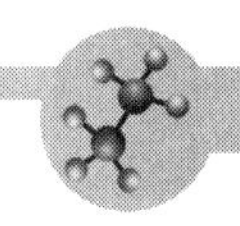

Dougherty's plants. The tobacco etch virus had created double-stranded RNA of its entire viral genome as it reproduced, as happens with many viruses. The plant cells responded by cutting off the RNA messages of all genes associated with the virus, including the CP genes incorporated into the plant DNA.

Biologists were stunned that such a powerful and ubiquitous system for regulating gene expression had escaped their notice for so long. Now that the shroud had been lifted on the phenomenon, scientists were anxious to analyze its mechanism of action and put it to gainful employment.

Slicing and Dicing Genetic Messages

RNA interference was soon observed in algae, flatworms and fruit flies—diverse branches of the evolutionary tree. Demonstrating RNAi within typical cells of humans and other mammals was considerably trickier, however.

When a human cell is infected by viruses that make long double-stranded RNAs, it can slam into lockdown mode: an enzyme known as PKR blocks translation of all messenger RNAs—both normal and viral—and the enzyme RNAse L indiscriminately destroys the messenger RNAs. These responses to double-stranded RNA are considered components of the so-called interferon response because they are triggered more readily after the cells have been exposed to interferons, molecules that infected cells secrete to signal danger to neighboring cells.

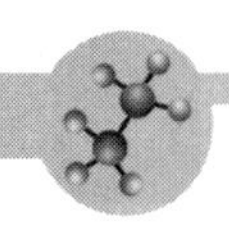

Unfortunately, when researchers put artificial double-stranded RNAs (like those used to induce RNA interference in worms and flies) into the cells of mature mammals, the interferon response indiscriminately shuts down every gene in the cell. A deeper understanding of how RNA interference works was needed before it could be used routinely without setting off the interferon alarms. In addition to the pioneering researchers already mentioned, Thomas Tuschl of the Rockefeller University, Phillip D. Zamore of the University of Massachusetts Medical School, Gregory Hannon of Cold Spring Harbor Laboratory in New York State and many others have added to our current understanding of the RNA interference mechanism.

RNAi appears to work like this: Inside a cell, double-stranded RNA encounters an enzyme dubbed Dicer. Using the chemical process of hydrolysis, Dicer cleaves the long RNA into pieces, known as short (or small) interfering RNAs, or siRNAs. Each siRNA is about 22 nucleotides long.

Dicer cuts through both strands of the long double-stranded RNA at slightly staggered positions so that each resulting siRNA has two overhanging nucleotides on one strand at either end. The siRNA duplex is then unwound, and one strand of the duplex is loaded into an assembly of proteins to form the RNA-induced silencing complex (RISC).

Within the silencing complex, the siRNA molecule is positioned so that messenger RNAs can bump into it.

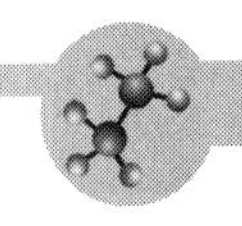

The RISC will encounter thousands of different messenger RNAs that are in a typical cell at any given moment. But the siRNA of the RISC will adhere well only to a messenger RNA that closely complements its own nucleotide sequence. So, unlike the interferon response, the silencing complex is highly selective in choosing its target messenger RNAs.

When a matched messenger RNA finally docks onto the siRNA, an enzyme known as Slicer cuts the captured messenger RNA strand in two. The RISC then releases the two messenger RNA pieces (now rendered incapable of directing protein synthesis) and moves on. The RISC itself stays intact, free to find and cleave another messenger RNA. In this way, the RNAi censor uses bits of the double-stranded RNA as a blacklist to identify and mute corresponding messenger RNAs.

David C. Baulcombe and his co-workers at the Sainsbury Laboratory in Norwich, England, were the first to spot siRNAs, in plants. Tuschl's group later isolated them from fruit fly embryos and demonstrated their role in gene silencing by synthesizing artificial siRNAs and using them to direct the destruction of messenger RNA targets. When that succeeded, Tuschl wondered whether these short snippets of RNA might slip under the radar of mammalian cells without setting off the interferon response, which generally ignores double-stranded RNAs that are shorter than 30 nucleotide pairs. He and his co-workers put synthetic siRNAs into cultured mammalian

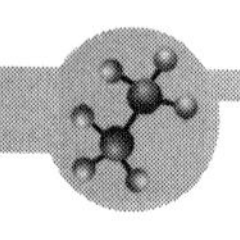

cells, and the experiment went just as they expected. The target genes were silenced; the interferon response never occurred.

Tuschl's findings rocked the biomedical community. Geneticists had long been able to introduce a new gene into mammalian cells by, for example, using viruses to ferry the gene into cells. But it would take labs months of labor to knock out a gene of interest to ascertain the gene's function. Now the dream of easily silencing a single, selected gene in mammalian cells was suddenly attainable. With siRNAs, almost any gene of interest can be turned off in mammalian cell cultures—including human cell lines—within a matter of hours. And the effect persists for days, long enough to complete an experiment.

A Dream Tool

As helpful as RNA interference has become to mammal biologists, it is even more useful at the moment to those who study lower organisms. A particular bonus for those studying worms and plants is that in these organisms the censorship effect is amplified and spread far from the site where the double-stranded RNA was introduced. This systemic phenomenon has allowed biologists to exploit RNAi in worms simply by feeding them bacteria engineered to make double-stranded RNA corresponding to the gene that should be shut down.

Because RNA interference is so easy to induce and yet so powerful, scientists are thinking big. Now that

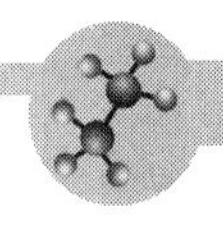

complete genomes—all the genes in the DNA—have been sequenced for a variety of organisms, scientists can use RNA interference to explore systematically what each gene does by turning it off. Recently four groups did just that in thousands of parallel experiments, each disabling a different gene of *C. elegans*. A similar genome-wide study is under way in plants, and several consortia are planning large RNAi studies of mammalian cells.

RNA interference is being used by pharmaceutical companies as well. Some drug designers are exploiting the effect as a shortcut to screen all genes of a certain kind in search of promising targets for new medicines. For instance, the systematic silencing of genes using RNAi could allow scientists to find a gene that is critical for the growth of certain cancer cells but not so important for the growth of normal cells. They could then develop a drug candidate that interferes with the protein product of this gene and then test the compound against cancer. Biotech firms have also been founded on the bet that gene silencing by RNAi could itself become a viable therapy to treat cancer, viral infections, certain dominant genetic disorders and other diseases that could be controlled by preventing selected genes from giving rise to illness-causing proteins.

Numerous reports have hinted at the promise of siRNAs for therapy. At least six labs have temporarily stopped viruses—HIV, polio and hepatitis C among them—from proliferating in human cell cultures. In each case, the scientists exposed the cells to siRNAs that

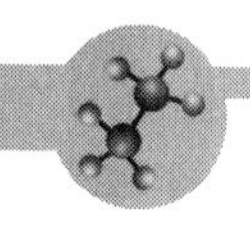

prompted cells to shut down production of proteins crucial to the pathogens' reproduction. More recently, groups led by Judy Lieberman of Harvard Medical School and Mark A. Kay of the Stanford University School of Medicine have reported that siRNAs injected under extremely high pressure into mice slowed hepatitis and rescued many of the animals from liver disease that otherwise would have killed them.

Despite these laboratory successes, it will be years before RNAi-based therapies can be used in hospitals. The most difficult challenge will probably be delivery. Although the RNAi effect can spread throughout a plant or worm, such spreading does not seem to occur in humans and other mammals. Also, siRNAs are very large compared with typical drugs and cannot be taken as pills, because the digestive tract will destroy them rather then absorb them. Researchers are testing various ways to disseminate siRNAs to many organs and to guide them through cells' outer membranes. But it is not yet clear whether any of the current strategies will work.

Another approach for solving the delivery problem is gene therapy. A novel gene that produces a particular siRNA might be loaded into a benign virus that will then bring the gene into the cells it infects. Beverly Davidson's group at the University of Iowa, for example, has used a modified adenovirus to deliver genes that produce siRNAs to the brain and liver of mice. Gene therapy in humans faces technical and regulatory difficulties, however.

Regardless of concerns about delivery, RNAi approaches have generated an excitement not currently

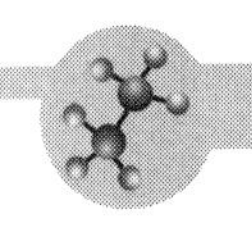

seen for antisense and catalytic RNA techniques—other methods that, in principle, could treat disease by impeding harmful messenger RNAs. This excitement stems in part from the realization that RNA interference harnesses natural gene-censoring machinery that evolution has perfected over time.

Why Cells Have Censors

Indeed the gene-censoring mechanism is thought to have emerged about a billion years ago to protect some common ancestor to plants, animals and fungi against viruses and mobile genetic elements. Supporting this idea, the groups of Ronald H. A. Plasterk at the Netherlands Cancer Institute and of Hervé Vaucheret at the French National Institute of Agricultural Research have shown that modern worms rely on RNA interference for protection against mobile genetic elements and that plants need it as a defense against viruses.

Yet RNA interference seems to play other biological roles as well. Mutant worms and weeds having an impaired Dicer enzyme or too little of it suffer from numerous developmental defects and cannot reproduce. Why should a Dicer deficiency cause animals and plants to look misshapen?

One hypothesis is that once nature developed such an effective mechanism for silencing the subversive genes in viruses and mobile DNA sequences, it started borrowing tools from the RNAi tool chest and using them for different purposes. Each cell has the same set of genes—what makes them different from one another

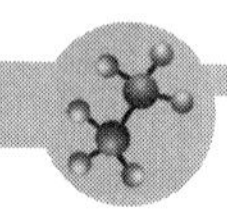

is which genes are expressed and which ones are not. Most plants and animals start as a single embryonic cell that divides and eventually gives rise to a multitude of cells of various types. For this to occur, many of the genes expressed in the embryonic cells need to be turned off as the organ matures. Other genes that are off need to be turned on. When the RNAi machinery is not defending against attack, it apparently pitches in to help silence normal cellular genes during developmental transitions needed to form disparate cell types, such as neurons and muscle cells, or different organs, such as the brain and heart.

What then motivates the RNAi machinery to hush particular normal genes within the cell? In some cases, a cell may naturally produce long double-stranded RNA specifically for this purpose. But frequently the triggers are "microRNAs"—small RNA fragments that resemble siRNAs but differ in origin. Whereas siRNAs come from the same types of genes or genomic regions that ultimately become silenced, microRNAs come from genes whose sole mission is to produce these tiny regulatory RNAs.

The RNA molecule initially transcribed from a microRNA gene—the microRNA precursor—folds back on itself, forming a structure that resembles an old-fashioned hairpin. With the help of Dicer, the middle section is chopped out of the hairpin, and the resulting piece typically behaves very much like an siRNA—with the important exception that it does not censor a gene with any resemblance to the one that produced it but instead censors some other gene altogether.

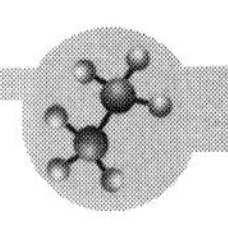

As with the RNAi phenomenon in general, it has taken biologists time to appreciate the potential of microRNAs for regulating gene expression. Until recently, scientists knew of only two microRNAs, called *lin-4* RNA and *let-7* RNA, discovered by the groups of Victor Ambros of Dartmouth Medical School and Gary Ruvkun of Harvard Medical School. In the past two years we, Tuschl, Ambros and others have discovered hundreds of additional microRNA genes in worms, flies, plants and humans.

With Christopher Burge at M.I.T., we have estimated that humans have between 200 and 255 microRNA genes—nearly 1 percent of the total number of human genes. The microRNA genes had escaped detection because the computer programs designed to sift through the reams of genomic sequence data had not been trained to find this unusual type of gene, whose final product is an RNA rather than a protein.

Some microRNAs, particularly those in plants, guide the slicing of their mRNA targets, as was shown by James C. Carrington of Oregon State University and Zamore. We and Bonnie Bartel of Rice University have noted that plant microRNAs take aim primarily at genes important for development. By clearing their messages from certain cells during development, RNAi could help the cells mature into the correct type and form the proper structures.

Interestingly, the *lin-4* and *let-7* RNAs, first discovered in worms because of their crucial role in pacing development, can employ a second tactic as well. The messenger RNAs targeted by these microRNAs are only approximately complementary to the microRNAs,

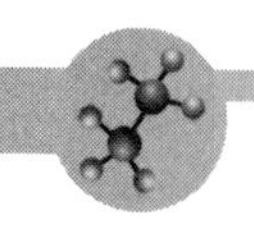

and these messages are not cleaved. Some other mechanism blocks translation of the messenger RNAs into productive proteins.

Faced with these different silencing mechanisms, biologists are keeping open minds about the roles of small RNAs and the RNAi machinery. Mounting evidence indicates that siRNAs not only capture messenger RNAs for destruction but can also direct the silencing of DNA—in the most extreme case, by literally editing genes right out of the genome. In most cases, however, the silenced DNA is not destroyed; instead it is more tightly packed so that it cannot be transcribed.

From its humble beginnings in white flowers and deformed worms, our understanding of RNA interference has come a long way. Almost all facets of biology, biomedicine and bioengineering are being touched by RNAi, as the gene-silencing technique spreads to more labs and experimental organisms.

Still, RNAi poses many fascinating questions. What is the span of biological processes that RNA interference, siRNAs and microRNAs influence? How does the RNAi molecular machinery operate at the level of atoms and chemical bonds? Do any diseases result from defects in the RNAi process and in microRNAs? As these questions yield to science, our understanding of the phenomenon will gradually solidify—perhaps into a foundation for an entirely new pillar of genetic medicine.

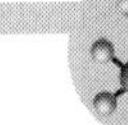

After fertilization, an organism starts as one round cell. Yet most embryos develop into organisms that have patterns such as a head end and a tail end. How does this happen? One idea is that gradients of chemicals called morphogens play an important role in development. Gradients are differences in concentrations of the morphogen across the embryo, and the gradient can establish an axis (head from tail). In the following article, Christiane Nüsslein-Volhard discusses experiments that show how morphogen gradients cause pattern formation in the Drosophila *embryo. —CCF*

"Gradients That Organize Embryo Development"

by Christiane Nüsslein-Volhard
***Scientific American*, August 1996**

A few crucial molecular signals give rise to chemical gradients that organize the developing embryo

Bears mate in wintertime. The female then retires into a cave to give birth, after several months, to three or four youngsters. At the time of birth, these are shapeless balls of flesh, only the claws are developed. The mother licks them into shape.

This ancient theory, recounted by Pliny the Elder, is one of the many bizarre early attempts to explain one of life's greatest mysteries—how a nearly uniform egg cell develops into an animal with dozens of types of cells, each in its proper place. The difficulty is finding an

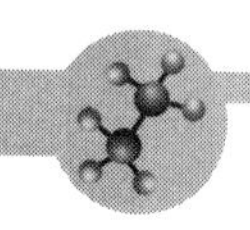

explanation for the striking increase in complexity. A more serious theory, popular in the 18th and 19th centuries, postulated that an egg cell is not structureless, as it appears, but contains an invisible mosaic of "determinants" that has only to unfold to give rise to the mature organism. It is hard for us now to understand how this idea could have been believed for such a long time. To contain the complete structure of the adult animal in invisible form, an egg would also have to contain the structures of all successive generations, because adult females will in time produce their own eggs, and so on, ad infinitum. Even Goethe, the great poet and naturalist, favored this "preformation hypothesis," because he could not think of any other explanation.

About 100 years ago experimental embryologists began to realize that developmental pathways need not be completely determined by the time the egg is formed. They discovered that some experimental manipulations led to dramatic changes in development that could not be explained by the mosaic hypothesis. If an experimenter splits a sea-urchin embryo at the two-cell stage into two single cells, for example, each of the cells will develop into a complete animal, even though together the two cells would have produced only one animal if left undisturbed. When human embryos split naturally, the result is identical twins.

Slowly an important idea emerged: the gradient hypothesis. One of the proposers of this idea was Theodor H. Boveri of the University of Würzburg, the founder of the chromosomal theory of inheritance. Boveri suggested that "a something increases or

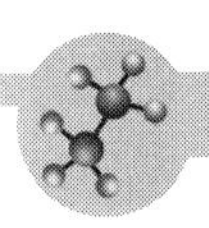

decreases in concentration" from one end of an egg to the other. The hypothesis, in essence, is that cells in a developing field respond to a special substance—a morphogen—the concentration of which gradually increases in a certain direction, forming a gradient. Different concentrations of the morphogen were postulated to cause different responses in cells.

Although concentration gradients of morphogens could in principle explain how cells "know" their position in an embryo, the idea was for a long time not widely accepted. One of the difficulties lay in explaining how a morphogenetic gradient could be established and then remain stable over a sufficient period. In a developing tissue composed of many cells, cell membranes would prevent the spread of large molecules that might form a concentration gradient. In a single large egg cell, conversely, diffusion would quickly level such a gradient. Further, the biochemical nature and the mechanism of action of morphogens were a mystery.

For most biologists, the means of gradient formation remained elusive until recently, when researchers in several laboratories discovered gradients operating in the early embryo of the fruit fly, *Drosophila*. For most nonbiologists, it is a surprise that many of the mechanisms of development are best known in *Drosophila*, rather than in animals more closely related to humans. The examples I shall describe illustrate the reason for the preeminence of *Drosophila* as an experimental subject: a lucky coincidence of advantages makes it almost ideal for studies in genetics, embryology and molecular biology.

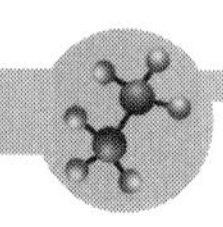

Drosophila became the laboratory animal of choice for studying Mendelian genetics early this century because the fly is easy to handle and quick to breed in large numbers, making it possible to search through many individual flies for mutants. Studies of mutants have successfully elucidated metabolic pathways and regulatory processes in viruses, bacteria and yeast. Twenty years ago Eric F. Wieschaus, now at Princeton University, and I extended this approach to *Drosophila* by searching for genes that control the segmented form of the larva. The larva is relatively large—about one millimeter long—and has well-defined, repeated segments that emerge within 24 hours of the laying of the egg. These features are crucial for interpreting experimentally induced abnormalities that affect the pattern of development.

Another key advantage of using *Drosophila* for embryological studies is that during its early development the embryo is not partitioned into separate cells. In the embryos of most animals, when a cell's nucleus divides, the rest of the cell contents divides with it. Cell membranes then segregate the halves, yielding two cells where there was one. Hence, the embryo grows as a ball of cells. In contrast, the nucleus of the fertilized *Drosophila* egg divides repeatedly, but membranes do not isolate the copies. Eventually thousands of nuclei lie around the periphery of what is still, in a manner of speaking, a single cell. Only after three hours of cell division, when some 6,000 nuclei have formed, do separating membranes appear.

This peculiarity allows chemicals to spread freely through the early embryo and influence the developmental fate of large regions of it. As experimentalists, we can

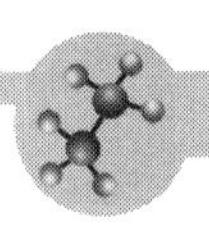

therefore transplant cytoplasm (the viscous fluid within cells) or inject biological molecules into various regions of a *Drosophila* embryo and observe the results.

The Power of Gradients

In addition, *Drosophila* is fairly easy to study with the techniques of molecular biology. The insect has only four pairs of chromosomes, and they exist in a special giant form. The giant chromosomes make it possible to see under the microscope, in many cases, the disruptions in the genetic material caused by mutations. This fact helps a great deal when the mutations are being studied. Last but not least, by exploiting naturally occurring mobile genetic elements, it is possible to add, with high efficiency, specific genes to the genetic complement of *Drosophila*.

By studying mutants, researchers have found about 30 genes that are active in the female and organize the pattern of the embryo. Only three of them encode molecular signals that specify the structures along the long antero-posterior (head-tail) axis of the larva. Each signal is located at a particular site in the developing egg and initiates the creation of a different type of morphogenetic gradient. In each case, the morphogen has its maximum concentration at the site of the signal.

One of the signals controls the development of the front half of the egg, which gives rise to the head and thorax of the larva. A second signal controls the region that develops into the abdomen, and the third controls development of structures at both extreme ends of the larva.

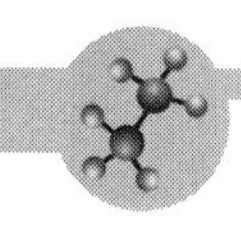

The simplest of the morphogenetic gradients initiated by these signals consists of a protein called Bicoid, which determines the pattern in the front part of the larva. My colleague Wolfgang Driever and I found that a concentration gradient of Bicoid is present in the *Drosophila* embryo from the very earliest stages. The concentration is highest at the head end of the embryo, and it declines gradually along the embryo's length. Mutations in the *bicoid* gene of a *Drosophila* female prevent the development of a Bicoid gradient. The resulting embryos lack a head and thorax.

Bicoid acts in the nuclei of the embryo. The protein is termed a transcription factor, because it can initiate transcription of a gene. This is the process whereby messenger RNA (mRNA) is produced from the genetic material, DNA; the cell then uses the mRNA to synthesize the gene's protein product. Transcription factors operate by binding to specific DNA sequences in the control regions, or promoters, of target genes. In order to bind to a promoter, Bicoid must be present above a certain threshold concentration.

Driever and I have investigated the interaction of Bicoid with one target gene in particular, *hunchback*. *Hunchback* is transcribed in the front half of the early embryo, and the gene's promoter contains several Bicoid binding sites. We carried out two types of experiment: in one, we changed the concentration profile of Bicoid, and in the other we changed the structure of the *hunchback* gene promoter.

By introducing additional copies of the *bicoid* gene into the female, it is possible to obtain eggs with levels of

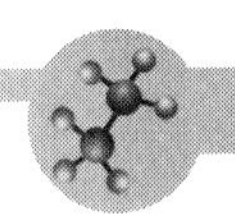

Bicoid that are four times higher than normal all along the gradient. In these embryos, the zone of *hunchback* gene activation extends toward the posterior, and the head and thorax develop from a larger part of the embryo than is normal. This abnormality could in principle arise either because the Bicoid concentration gradient was steeper in the manipulated embryos or because the absolute level of Bicoid concentration was higher. The correct interpretation was made clear by an experiment in which we made mutant embryos that had a level Bicoid concentration along their length, so there was no gradient at all. These embryos produced only one type of anterior structure (head or thorax); which type depended on the Bicoid concentration. The experiment shows, then, that the absolute concentration of Bicoid, not the steepness of the gradient, is important for controlling subsequent development of each region.

In the second type of experiment the Bicoid gradient was left unchanged, but the promoter region of the *hunchback* gene was altered. When the altered promoter bound only weakly to Bicoid, higher Bicoid concentrations were required to initiate *hunchback* transcription. Consequently, the edge of the zone of *hunchback* activity shifted forward. In these embryos, as one might predict, the head forms from a smaller than normal region. This experiment revealed that Bicoid exerts its effect by binding to the *hunchback* promoter.

These experiments show how a morphogen such as Bicoid can specify the position of a gene's activation in an embryo through its affinity for the gene, in this case *hunchback*. In theory, a large number of target genes could respond to various thresholds within the gradient of a single morphogen, producing many different zones of gene activation. In reality, however, a gradient acts directly on usually no more than two or three genes, so it specifies only two or three zones of activation.

How is the morphogenetic Bicoid gradient itself established? While the unfertilized egg is developing, special nurse cells connected to it deposit mRNA for Bicoid at its anterior tip. Synthesis of Bicoid, which starts at fertilization, is therefore already under way when the egg is laid. As the embryo develops, the protein diffuses away from the site of its production at the head end. Bicoid is unstable, however, so its concentrations at remote points—that is, at the end that will become the abdomen—never reach high levels. The resulting concentration gradient persists until cell membranes form.

This simple diffusion mechanism is accurate enough to meet the requirements of normal development. Remarkably, even substantial changes in Bicoid levels—doubling or halving—result in normally proportioned larvae. It appears that mechanisms operating at later stages of development can correct some errors in the early stages. If a researcher transplants *bicoid* mRNA into the posterior pole of a normal embryo, an additional Bicoid protein gradient arises, oriented opposite to the natural one. The resulting embryo displays a duplicate head where the abdomen should be. This experiment shows conclusively that *bicoid* mRNA is by itself sufficient to determine polarity.

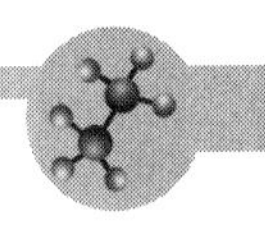

Other work has revealed how the *bicoid* mRNA is positioned precisely within the egg cell. Paul M. Macdonald of Stanford University has identified a large section of the *bicoid* mRNA molecule that contains all the information required for a cell to recognize it, transport it and anchor it. Daniel St. Johnston and Dominique Ferrandon, while working in my laboratory, found that a molecular complex consisting of *bicoid* mRNA and a protein known as Staufen will move in one direction along structural elements in cells called microtubules. It seems likely that this effect explains the localization of *bicoid* mRNA, although other proteins are certainly also involved.

Whereas Bicoid is determining the anterior section of the larva's long axis, a different morphogenetic gradient is determining the posterior part. The gradient in this case is composed of the protein Nanos. *Nanos* mRNA localizes in the cytoplasm at the posterior end of the egg. This occurrence depends critically on another molecular complex consisting of the Staufen protein and mRNA from a gene named *oskar*. Anne Ephrussi and Ruth Lehmann, then at the Whitehead Institute for Biomedical Research in Cambridge, Mass., demonstrated the crucial role of *oskar* by replacing the section of mRNA required for localization with that section of *bicoid* mRNA. This hybrid molecule behaved like *bicoid* mRNA, collecting at the anterior pole rather than at the posterior one. The manipulation misdirected the *nanos* mRNA to the anterior pole, causing the embryos to develop with two abdominal ends in mirror symmetry.

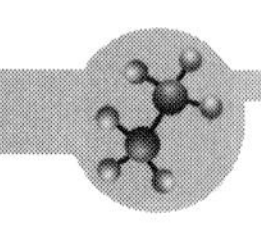

Getting Around Cell Membranes

The mechanisms that produce the morphogenetic gradients of Bicoid and Nanos, both of which are large molecules, can operate only when there are no cell membranes to hinder diffusion. In most animals, however, early development creates cell membranes between different regions of the egg, so these mechanisms cannot work. It is notable, then, that the dorsoventral (top-bottom) axis of the *Drosophila* embryo, unlike the antero-posterior axis, is defined by a single gradient that could develop even in the presence of cell membranes. This mechanism may thus be more typical of those found in other creatures.

The first embryonic pattern along the dorsoventral axis is determined by the gradient of a protein called Dorsal. Like Bicoid, Dorsal is a transcription factor, and it controls the activity of several target genes in a concentration-dependent manner. The Dorsal protein acts as both a transcriptional activator and a repressor—inside cell nuclei, it turns genes on or off. When its concentration in the cell nucleus exceeds a particular threshold, Dorsal activates the transcription of a pair of genes that play important roles in subsequent development. Whenever Dorsal's nuclear concentration exceeds a lower threshold, it *represses* the transcription of two different genes. If the concentration of Dorsal in the various cell nuclei is arranged as a gradient, each of these pairs of genes will subsequently be expressed on a different side of the embryo.

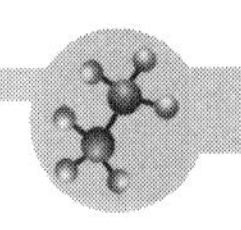

The formation of the nuclear concentration gradient of Dorsal protein is, however, entirely different from the formation of the Bicoid gradient. Overall, the concentration of Dorsal protein is actually level throughout the embryo. Christine W. Rushlow and Michael S. Levine of Columbia University, along with my colleague Siegfried Roth and me, have shown that what does vary along the dorsoventral axis of the embryo is the degree to which Dorsal protein is sequestered in nuclei. Close to the dorsal side of the embryo, the protein is found increasingly within the cytoplasm; on the ventral side it is found mainly within nuclei.

How does this strange gradient of Dorsal concentrated in nuclei arise? Normally, what stops Dorsal from entering nuclei is a protein called Cactus which binds to it. On the ventral side of the embryo, however, Dorsal is released from this bound state by an activation pathway involving at least 10 proteins.

The ventral signal that starts this process originates early in egg development inside the female. Yet its effect—the importation of Dorsal to the nucleus—takes place several hours later, in embryos with rapidly dividing nuclei. Thus, the signal must be very stable. The signal's exact nature remains unclear, but it is concentrated in the specialized membrane—known as the vitelline membrane—that surrounds the egg after it is laid.

Painstaking experiments by my colleague David Stein and me and by Kathryn V. Anderson and her colleagues at the University of California at Berkeley have

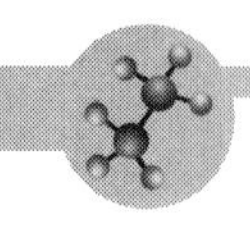

established that some early components of the activation pathway are produced in the mother's follicle cells, which surround the unlaid egg. Others are produced in the egg cell and then deposited either in the egg's cytoplasm or in its cell membrane or secreted into the space surrounding the egg.

Initially, the protein components of this pathway are evenly distributed, each in its proper compartment. Then the signal, which identifies the ventral side, becomes active. This signal seems to determine the Dorsal gradient by triggering a cascade of interactions among the proteins of the activation pathway; the cascade conveys into the egg the information about which side will be ventral.

This message relay system probably relies on gradients of its own. It seems likely that a true gradient first appears in the space surrounding the egg cell, because large proteins can easily diffuse through this region. The gradient signal is thought to cause graded activation of a receptor molecule in the egg's cell membrane; that is, the receptors may become either more or less active depending on how ventral their position is. The receptors could then transmit a similarly graded signal into the egg cytoplasm, and so on.

Thus, the signal that initiates the formation of the embryo's dorsoventral pattern circumvents the obstacle to diffusion. In order to do this, it relies on a message relay system that, through a variety of protein molecules, carries the gradient information from one compartment to another. (A similar mechanism for carrying a signal across the egg cell membrane operates in the terminal

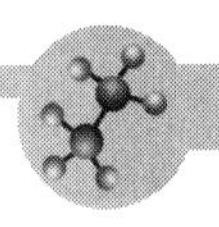

pathway, which is the system that controls structures at both ends of the antero-posterior axis.) In this manner, signals from outside an egg, where a gradient can easily form by diffusion, can be transmitted to the inside. The result is the graded importation into the nuclei of a protein that was initially evenly distributed.

Patterns in Common

What conclusions can we draw from these investigations? Before gradients were identified, biologists believed that morphogens might constitute a special class of molecule with unique properties. This is clearly not the case. In the early *Drosophila* embryo, many "ordinary" proteins that can serve different biochemical functions can convey positional information.

In some instances, such as the process determining the dorsoventral pattern, a gradient arises first by diffusion and is then copied down a molecular chain of command by activation of successive proteins. In other cases, gradients have inhibitory effects. The Nanos gradient, for example, represses the cell's use of one type of evenly distributed mRNA, thereby creating a gradient of the opposite orientation.

In all the pathways so far investigated, the final result is a gradient of a morphogen that functions principally as a transcription factor, initiating or suppressing the transcription of one or more target genes in a concentration-dependent manner. These gradients are sometimes quite shallow: Bicoid and Dorsal decline in concentration only slowly along the length of the embryo. Yet they somehow cause the protein

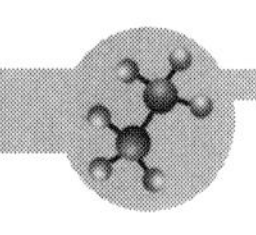

products of their target genes to have sharp cutoff points. How can this happen?

One way this might occur is if several molecules—either different ones or multiple copies of the same one—cooperated to bring about transcription. The dynamics often result in a steep dependence on the concentration of one or more of the components. It is noteworthy, then, that genes activated by Bicoid or Dorsal proteins contain multiple adjacent binding sites, often for different transcription factors that may modulate the genes' activity.

Some morphogenetic gradients apparently yield but a single effect: if the concentration of the morphogen in a particular place is above a critical threshold, a target gene is activated; otherwise, it is not. In other cases, different concentrations of morphogen elicit different responses, and it is this type of gradient that is most important for providing an increase in the complexity of the developing organism.

Although each morphogenetic gradient seems to control only a few target genes directly, interactions between cofactor molecules that affect transcription can radically change responses to the gradients. These mechanisms of combinatorial regulation open the way to the formation of patterns of great complexity from an initially simple system. Proteins acting as cofactors can modify a morphogen's affinity for a gene's promoter region, thus shifting a critical threshold up or down. A cofactor might even turn an activating transcription factor into a repressor. The potential for creating complex patterns becomes

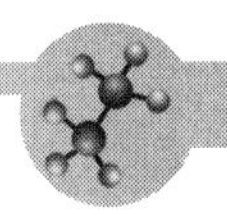

apparent when one considers that the cofactors may themselves be distributed in a graded fashion.

Superposing several gradients onto an embryonic region can subdivide it even more and generate additional complexity. The three pathways that define the anteroposterior axis of the *Drosophila* embryo together give rise to four separate and independent gradients (the terminal pathway produces two gradients, of an unknown protein). Each gradient has one or two thresholds. At least seven regions are thus defined by a unique combination of target gene expression. At the anterior end, where the gradient of the as yet unidentified terminal protein and the Bicoid gradient overlap, the combination leads to the development of the foremost extreme of *Drosophila*, a part of the head. The gradient of the unknown protein acting alone, in contrast, produces the structures of the opposite end, at the tip of the abdomen.

Combinatorial regulation as a principle of pattern formation is even more apparent later in fruit-fly development. For example, the gradients of transcription factors along the long axis of the embryo affect genes that, in most cases, encode other transcription factors. Those secondary factors, in turn, diffuse out into gradients of their own. At various threshold concentrations, each factor acts on its own gene targets; sometimes these thresholds are altered by other transcription factors with overlapping spheres of influence.

Concentration dependence and combinatorial regulation together open up a versatile repertoire of pattern-forming mechanisms that can realize the designs

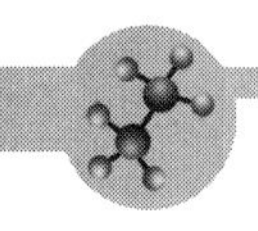

encoded in genes. In *Drosophila*, the initial patterns generate transverse stripes of gene expression covering the part of the egg to be segmented in the larva.

This pattern in turn directs the formation of an even more finely striped pattern, which then directly determines the characteristics of each segment in the embryo. As soon as the embryo partitions itself into cells, transcription factors can no longer diffuse through the cell layers. The later steps of pattern refinement therefore rely on signaling between neighboring cells, probably with special mechanisms carrying signals across cell membranes.

Many more details remain to be discovered before we have a complete picture of how the *Drosophila* embryo develops. Yet I believe we have now uncovered some of the principal features. This accomplishment can illuminate much of zoology, because one great surprise of the past five years has been the discovery that very similar basic mechanisms, involving similar genes and transcription factors, operate in early development throughout the animal kingdom.

Basic research on a good model system has thus led to powerful insights that might one day help us understand human development. What these insights have already provided is a satisfying answer to one of the most profound questions in nature—how complexity arises from initial simplicity.

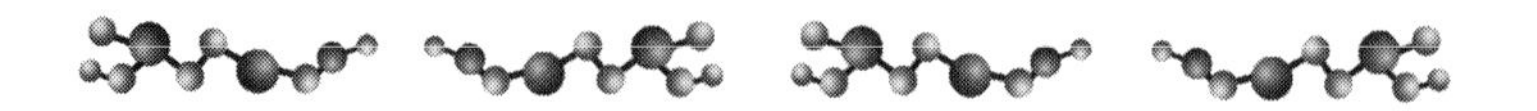

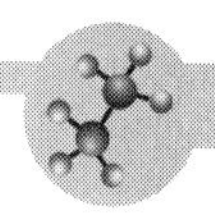

Morphogen gradients such as Bicoid and Nanos are important in pattern formation in the Drosophila embryo, where the embryo remains a single cell in early development. But what about other organisms where this situation does not occur? Are morphogenic gradients established in the same ways? In the following article, Robert D. Riddle and Clifford J. Tabin describe experiments on limb development in chickens and mice that establish how various cells and the expression of the "hedgehog" family of genes are crucial to development in general. —CCF

"How Limbs Develop"
by Robert D. Riddle and Clifford J. Tabin
***Scientific American*, February 1999**

A protein playfully named Sonic hedgehog is one of the long-sought factors that dictate the pattern of limb development

The waiting was the hardest part. But finally it was time to crack some eggs. A week earlier we had cut a small hole in the eggshell of a developing chick and inserted some genetically engineered cells into one of the two tiny buds that were destined to develop into the embryo's wings. We had engineered the cells to make a protein that we suspected to be one of the major determinants in establishing the overall pattern of wing development. Now was the moment of truth: How had the extra cells affected the formation of the limb?

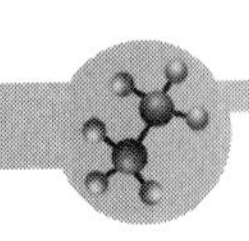

As we peered into the microscope to examine the embryo closely, we saw that our highest hopes had been realized. The transplanted cells had caused a whole new set of digits to form at the wing tip, confirming that we had identified an important factor in limb development.

This experiment, which we conducted in the summer of 1993, partially answered a question that biologists had posed early this century: How do the cells in a developing limb "know" left from right, top from bottom, and back from front? More specifically, what caused the digit that faces forward (anterior) when you hold your arms at your sides to form a thumb, and what caused the digit that faces toward the back (posterior) to form a pinkie? What ensured that the bone of your upper arm formed close (proximal) to your body while your fingers took shape farther away (distal)? And why did only those cells on the bottom (ventral) sides of your hands form creased palms, with the back (dorsal) sides remaining smooth?

Experimental embryologists have been trying to answer these questions for decades. Until recently, however, most studies have focused on identifying the cells that are necessary for proper limb development. With the advent of molecular biology techniques, scientists can now analyze the specific genes that direct the formation of limbs. Curiously, many of the genes—and the proteins they make—are closely related to others that control the development of the limbs of fruit flies, even though vertebrates and insects are thought to have evolved from a common ancestor that lacked limbs altogether.

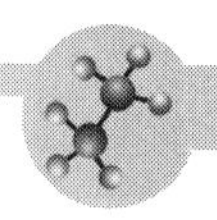

Besides satisfying age-old curiosities about the miracle of life, such studies are helping researchers to understand how and why the processes of embryonic development sometimes go wrong, resulting in birth defects. What is more, they are indicating that the same protein that establishes the anterior and posterior sides of developing limbs affects a number of other developmental processes, from the formation of the central nervous system to the growth of cells that can cause a form of skin cancer.

One of biology's oldest questions concerns whether all organisms share similar factors or processes that guide embryonic development or if each particular organism or group of organisms develops in a manner unique to it. It might seem obvious that human arms develop similarly to those of chimpanzees, for instance, but how similar are human and chimpanzee arms to chicken wings? And does the development of mammalian arms have anything in common with the formation of the wings of flies?

For years, biologists assumed that the factors that shape the developing legs of a future ballerina and those of a pesky fruit fly are very different. Any similarities between the two were thought to be simply the result of convergent evolution, in which similar structures arise through entirely different means. But two revolutionary ideas have now emerged to change that line of thinking.

First, biologists now know that the same or similar genes shape the development of many comparable structures across the spectrum of the animal kingdom,

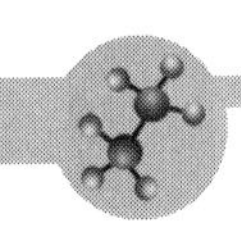

from flies to mice to humans. Nearly every animal has a head on one end of its body and a tail at the other end, for example, because of the activity of a family of genes called the homeobox, or Hox genes [see "The Molecular Architects of Body Design," by William McGinnis and Michael Kuziora; SCIENTIFIC AMERICAN, February 1994]. Second, genes that direct the formation of one aspect of development—for instance, the sculpting of limbs—can also play a role in something as different as the formation of the nervous system. Nature, it seems, uses the same toolbox again and again to put together amazingly diverse organisms.

Chicken eggs are particularly useful for studying how limbs form. Since the time of Aristotle, scientists have known that to observe how a chicken embryo develops, one simply needs to cut a hole in the shell of a fertilized egg. For well over 100 years, embryologists have surgically altered chicken embryos through such small holes, then sealed the holes with paraffin wax (or Scotch tape, today) and incubated the eggs until hatching.

Through such studies, researchers have observed that limbs form initially from buds that appear along the sides of the developing body. The buds consist of a "jacket" of outer cells, or ectoderm, surrounding a core of other cells called the mesoderm. Although early limb buds are not fully organized structures, they contain all the information required to form a limb: removing an early limb bud and transplanting it to another site on an embryo results in the growth of a normal limb in an abnormal location.

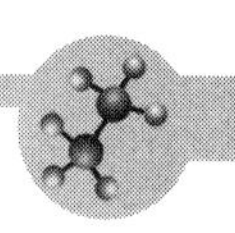

At these early stages of development, the cells in a limb bud are not "committed" to becoming part of a thumb or a pinkie—they are in the midst of the process of becoming one or the other. Accordingly, they can be poked and prodded to help experimenters understand the rules of limb development.

In every limb bud, there are leaders and there are followers. Through the years, developmental biologists have determined that each axis of a developing limb—anterior-posterior, proximal-distal and dorsal-ventral—is organized by distinct types of cells in distinct locations in the limb bud. These cells are referred to as signaling centers.

The ectoderm, for example, will make only a part of the skin of the adult, but it establishes the dorsal-ventral axis that affects the location and formation of each of the muscles and tendons. Scientists have known for years that removing the ectoderm from an early limb bud, rotating it 180 degrees and replacing it causes muscles that normally develop on the dorsal side of a chick wing to end up on the ventral side, and vice versa. The ectoderm accomplishes this by sending chemical signals to the underlying cells that will eventually form the muscles and tendons.

Beginning in the late 1940s, John W. Saunders, Jr., of the State University of New York at Albany and his colleagues observed that a particular clump of ectodermal cells at the tip of each developing chick limb bud—which they termed the apical ectodermal ridge (AER)—is responsible for setting up a limb's proximal-distal pattern. When they removed this

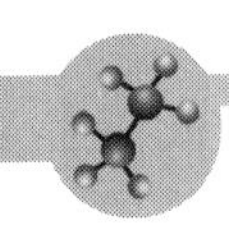

ridge of cells, only a stump of a limb developed; when they transplanted an extra clump of the cells onto an otherwise normal limb bud, it developed into a double limb. Moreover, the timing of the microsurgery determined how much of the limb would form and how far down the first limb the second limb would start growing. This demonstrated that the AER is both necessary and sufficient for limb outgrowth. Furthermore, because the limbs grew proximal structures first and distal structures later, the experiment showed that the AER regulates development along the proximal-distal axis.

Saunders and his co-workers also identified a second clump of cells that dictates the anterior-posterior axis of a budding limb. These cells lie just beneath the ectoderm, along the posterior edge of the limb bud. When the researchers transplanted the cells from the posterior side of one chick limb bud to the anterior side of a second bud, they found that the limb formed an entire set of additional digits—only oriented backward, so that the chicken equivalent of our little finger faced the front. Because the transplanted cells not only induced the anterior cells to form extra digits but also turned, or repolarized, them, Saunders's group labeled the region from which they took the cells the zone of polarizing activity (ZPA).

In the mid-1970s Cheryll Tickle, now at the University of Dundee, found that the ZPA works in a concentration-dependent manner: the more cells transplanted, the more digits duplicated. This evidence suggested that the ZPA functions by secreting a

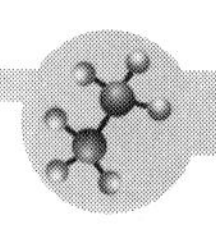

chemical signal called a morphogen that becomes fainter as it diffuses throughout the limb bud. (The idea of a morphogen was first proposed around the turn of the century and greatly expanded on in the late 1960s by Lewis Wolpert, who is now at University College London.) According to the morphogen hypothesis, anterior digits form from cells farthest from the ZPA, which are exposed to the lowest concentrations of the morphogen, and posterior digits form closer to the ZPA, where they are exposed to higher morphogen concentrations. But the identity of the morphogen and exactly how it functioned remained a mystery.

Shape of Things to Come

The advent of molecular biology has given researchers the means to identify the genes involved in embryonic processes such as limb development. Once a gene is cloned—that is, once its DNA is isolated—it becomes a powerful research tool. By reading the order of the chemical "letters" that make up the gene's DNA, scientists can predict the structure of the protein it encodes. Pieces of the DNA can also be used as probes to determine where the gene is active in a developing embryo and when it is turned on. Perhaps most important, once a gene is cloned, scientists can alter the gene's expression, switching it on in places where it normally does not function or shutting it off where it normally would be on. By doing so, researchers can begin to explore the function of the gene during normal development.

During the 1980s and 1990s, researchers studying fruit flies discovered that the posterior parts of the various segments that make up the embryonic fruit-fly body plan produce a protein that is vital for fly development. The protein was named hedgehog because fly larvae with mutations in the gene that encodes the protein do not develop normally but instead appear rolled up and bristly, like frightened hedgehogs.

To determine whether a similar protein might play a role in the development of vertebrates, we collaborated with researchers led by Andrew P. McMahon of Harvard University and Philip W. Ingham—now at the University of Sheffield—to use probes from the fly *hedgehog* gene to search for comparable genes in mice, chickens and fish. Between us, we turned up not one but three versions of the gene. We named them after three types of hedgehog: *Desert hedgehog*, for a species prevalent in North Africa; *Indian hedgehog*, after a variety indigenous to the Indian subcontinent; and *Sonic hedgehog*, for the Sega computer game character found in video arcades worldwide.

We discovered that all three genes exist in mice, chickens and fish but that each one has a different function in the development of those organisms. *Desert hedgehog*, for example, is important in sperm production because male mice with mutations in the gene are sterile. *Indian hedgehog* is expressed in growing cartilage, where it plays a role in cartilage development. But *Sonic hedgehog* has a truly remarkable pattern of expression that suggests it functions in the development of other body regions as well. Not only is *Sonic hedgehog*

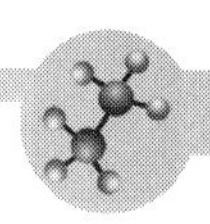

active in the ZPAs of limb buds, it is also "on" in a region of the developing spinal cord that acts as a signaling center in its own right. In addition, it is known to prompt the growth of extra digits when transplanted onto a budding limb.

Because other scientists had found that the fruit-fly hedgehog protein is secreted, we guessed that Sonic hedgehog might be one of the signals that shapes the growth of vertebrate limbs. To test this idea directly, we spliced the *Sonic hedgehog* gene into embryonic chick cells grown in the laboratory, causing the cells to produce the Sonic hedgehog protein, and then implanted the cells into the anterior side of a chick limb bud. Just as in Saunders's experiments, the transplanted cells prompted the formation of a duplicate set of digits that were oriented backward.

Since our 1993 studies, Tickle and McMahon have found that purified Sonic hedgehog protein has the same effect as the gene-spliced cells, proving that the protein is indeed responsible for establishing the anterior-posterior axis during vertebrate limb development. Moreover, Sonic hedgehog functions just as we would expect a morphogen to: high concentrations produce a full set of extra reverse-ordered digits, whereas low concentrations result in fewer duplicated, backward structures. Scientists are now studying the molecular nature of this concentration effect.

The early 1990s proved to be banner years for developmental biology. At about the same time that we and our collaborators were identifying Sonic hedgehog as the chemical signal that establishes the anterior-posterior axis of a developing limb, others were isolating the factors made by cells in the AER that set up the proximal-distal axis. Working independently, Tickle, Gail R. Martin of the University of California at San Francisco, John F. Fallon of the University of Wisconsin–Madison and their colleagues found that the AER makes several proteins called fibroblast growth factors that tell cells in a budding limb how far from the body to grow.

Tickle, Martin and Fallon determined that purified fibroblast growth factors could substitute for transplanted cells from the AER in driving limb outgrowth. They observed that normal limbs grew when they stapled tiny beads soaked in the factors to the tip of a limb bud that had had its AER removed. Such limb buds usually develop only severely shortened limbs.

We now know that the production of Sonic hedgehog and of the fibroblast growth factors is coordinated in a developing limb, allowing growth along the anterior-posterior axis to keep pace with that along the proximal-distal axis. Studies by our teams and by Lee A. Niswander of Memorial Sloan-Kettering Cancer Center in New York City have demonstrated that removing the fibroblast growth factor-producing AER from a chick limb bud shuts down the ability of that limb bud's ZPA to make Sonic hedgehog. Likewise, cutting out the ZPA prevents the AER from generating the fibroblast growth factors. But adding back the fibroblast growth factors allows Sonic hedgehog to be made. Similarly, reintroducing

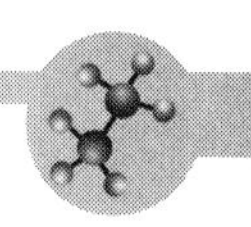

Sonic hedgehog fosters the production of the fibroblast growth factors.

Philip A. Beachy of the Johns Hopkins University School of Medicine and Heiner Westphal of the National Institute of Child Health and Human Development and their co-workers were the first to show that *Sonic hedgehog* is necessary for the proper functioning of the AER and ZPA in mice. They deleted the *Sonic hedgehog* gene to generate so-called knockout mice and saw that the mice developed severely shortened limbs that had failed to develop properly along both the anterior-posterior axis and the proximal-distal axis. Therefore, *Sonic hedgehog* is necessary and sufficient for normal limb development.

Birth Defects

The knockout mice have also indicated another dramatic role played by the Sonic hedgehog protein: generating the pattern in the brain and spinal cord that determines, for example, whether early neural cells become motor or sensory neurons. Besides developing extremely foreshortened limbs, mice lacking *Sonic hedgehog* have only one eye and have a severe brain defect called holoprosencephaly, in which the forebrain fails to divide into two lobes. Normal motor and sensory neuron development and the formation of two eyes and a bilateral brain depends on the activity of *Sonic hedgehog* in the neural tube—the precursor of the adult central nervous system—and in the cells beneath the tube.

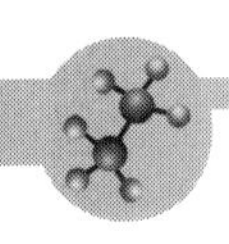

Holoprosencephaly is the most frequent congenital brain anomaly in the human forebrain. It can arise sporadically but also runs in families as part of several rare, inherited disorders. The degree of holoprosencephaly varies widely in affected individuals: some have mild cognitive deficits, whereas others have marked impairment accompanied by head and facial skeletal deformities.

Maximilian A. Muenke of the Children's Hospital of Philadelphia, Stephen W. Scherer of the Hospital for Sick Children in Toronto and their colleagues reported that mutations that inactivate *Sonic hedgehog* are responsible for some sporadic and inherited cases of holoprosencephaly. Without *Sonic hedgehog* to specify correctly the dorsalventral axis in the developing forebrain, the forebrain and eye tissues fail to become bilateral structures.

Of Hedgehogs and Cancer

It might not be surprising that a development-regulating gene such as *Sonic hedgehog* contributes to birth defects, but researchers have also recently uncovered a truly surprising link between the gene and cancer. The protein encoded by *Sonic hedgehog* signals cells by binding to the same protein on the cell surface that is involved in the skin cancer basal cell carcinoma.

Most chemical factors interact with susceptible cells by binding to cell-surface proteins called receptors. The binding of the factors to their specific receptors triggers a cascade of signals within the cell, ultimately leading to genes being turned on or off.

The receptors to which Sonic hedgehog binds consist of two subunits: one, called Smoothened, is poised to

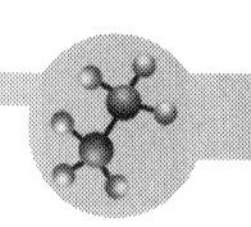

send a signal into the cell, and another, called Patched, keeps the first subunit from sending its signal. When Sonic hedgehog binds to Patched, it causes Patched to unleash Smoothened, the signaling subunit. In cells that harbor mutations that prevent them from making functional Patched proteins, however, the signaling half of the receptor is continuously active, as if Sonic hedgehog were constantly bathing the cells. Exactly how Sonic hedgehog affects normal skin development—and how the aberrant signaling of Smoothened leads to basal cell carcinoma—is now under intensive study.

Basal cell carcinoma is a malignancy of the epidermis or of skin cells lining the hair follicles that often results from mutations caused by overexposure to the ultraviolet radiation in sunlight. In 1996 Allen E. Bale of Yale University and Matthew P. Scott of Stanford University found independently that cancer arises when hair-follicle cells develop mutations in both copies of the *Patched* gene, the one inherited from the mother and the other handed down by the father: The mutations can occur in both copies of the gene after birth, or individuals can be born with a mutation in one copy, which makes them highly prone to developing multiple basal cell carcinomas if the second copy becomes mutated later.

Basal cell carcinoma is highly treatable but often recurs. If researchers could find small molecules to block the activity of Smoothened, the compounds might be used to prevent the cancers. Because such drugs could be applied directly to the skin, rather than taken orally, they might lack the side effects of systemic chemotherapies.

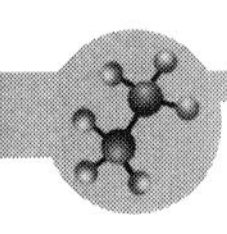

The role of Sonic hedgehog signaling in cancer should not be surprising. Molecular biology has shown in several instances that the processes that dictate development and cancer share many fundamental properties. The same factors that regulate cell growth and development in embryos, for example, also do so in adults. When mutations arise in the genes encoding these factors in embryos, birth defects occur; when they take place in adults, tumors can form.

Perhaps what is surprising is the degree to which a single factor like Sonic hedgehog can play various roles in the formation and function of an organism. Sonic hedgehog appears to be ancient: both flies and vertebrates have found multiple uses for it and many other embryonic genes. Once a molecular signaling pathway is established, nature often finds ways to use it in many other settings. One of the recurring themes of the symphony of life could be the sound of a Sonic hedgehog.

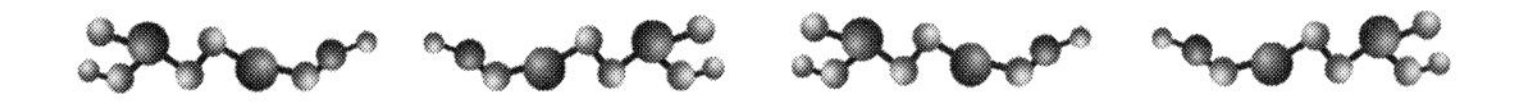

Many animals are bilaterally symmetric. However, the arrangement of internal organs is not symmetrical. Morphogen gradients appear to play a key role in this aspect of development as well. So far, we have discussed morphogens such as Bicoid, Nanos, and the hedgehog family, but are there other morphogens? Are morphogen

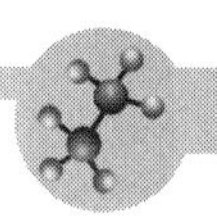

gradients the only components necessary for development? In the following article, Juan Carlos Izpisúa Belmonte discusses how asymmetry is produced in the developing heart. —CCF

"How the Body Tells Left from Right"
by Juan Carlos Izpisúa Belmonte
***Scientific American*, June 1999**

The precise orientation of our internal organs—and those of all other animals with a backbone—is controlled in part by proteins that are produced on only one side of an embryo

Look in the mirror and draw an imaginary line from the top of your head, along the bridge of your nose, and so on down your chest and your abdomen. You will notice that every external anatomical structure on one side of the line has a counterpart on the other side. Yet you have only one heart, one liver, one stomach, one pancreas and one spleen, and your colon coils from your right to your left. Even those organs that come in pairs show some asymmetry: the right lung has more lobes than the left, for instance, and some cerebral structures occur on only one side, or hemisphere, of the brain.

Why do our internal organs defy the symmetry of our overall body plan? And how do they get that way? Attempting to answer these questions, scientists have now identified several of the molecules that dictate organ placement, structure and orientation. We are finding that

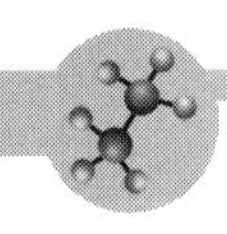

when these factors are absent or are produced in the wrong place, various human disorders can result. By understanding exactly how the factors function, we may learn how to treat or prevent such diseases.

A Place for Everything

Asymmetric organ structure and placement seem to have evolved because they offer advantages for survival. The very complex digestive system of higher vertebrates—organisms with a backbone—can be more efficiently packed in the body cavity, for example, if the system follows an asymmetric pattern of loops and turns. Similarly, an asymmetric heart is better able to pump and distribute blood. Such cardiac asymmetry allows for two separate pumping systems: one for directing blood to the lungs, where it can take up oxygen and discharge carbon dioxide, and a second for delivering the reoxygenated blood to the body.

Interestingly, internal organs can develop in a mirror-image fashion to the usual arrangement and still work properly. Approximately one in every 8,000 to 25,000 people is born with a condition known as situs inversus, in which the positions of all the internal organs are reversed relative to the normal situation (situs solitus): the person's heart and stomach lie to the right, their liver to the left, and so on. (The organs are also mirror images of their normal structures.) These people are usually healthy, suggesting that as long as all a person's organs turn and loop with a specific pattern or internal logic, the actual direction of turning and looping is not important.

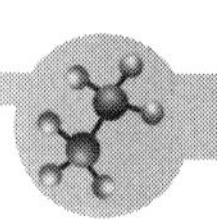

. . . And Everything in Its Place

People who are born with abnormally placed organs that are not a mirror image of the usual pattern are not so fortunate. Such individuals, who are said to have situs ambiguus, often die at an early age from lung or heart complications. Others who are born with a condition known as isomerism essentially have two left sides or two right sides to their bodies, so that they either have two spleens or no spleen, for instance. And internally, their hearts are exactly symmetric. The spectrum of disorders experienced by people with isomerism is complex, but for reasons that are still unclear those with left isomerism fare better than those with right isomerism. Many people with left isomerism have no symptoms at all, whereas those with right isomerism rarely survive beyond one year.

Researchers have elucidated some of the mechanisms that control left-right asymmetry by studying the early stages of heart development in embryos. They concentrate on the heart because it is the organ most sensitive to abnormalities in the biological machinery controlling the body plan.

All asymmetric organisms begin as symmetric embryos. As far as anyone can tell, all vertebrate embryos are perfectly bilaterally symmetric at the earliest stages of development, with the left side an apparently perfect mirror image of the right side. But at some point early on, this evenness is broken. In vertebrates, the first obvious indication is a very specific event during the initial stages of forming the heart.

The heart arises from two symmetric groups of precardiac cells (the so-called heart fields) that fuse as development progresses, forming an initially symmetric heart tube. The first visible asymmetry is the bending of this tube to the right. This "looping" of the heart tube is one of the most crucial steps in heart formation because it determines the internal structure of the two pumping systems.

In 1995 Clifford J. Tabin of Harvard Medical School, Claudio D. Stern of Columbia University and their colleagues identified one of the biochemical factors that induces looping of the developing heart tube [see "How Limbs Develop," by Robert D. Riddle and Clifford J. Tabin; SCIENTIFIC AMERICAN, February]. Studying chick embryos, these researchers found that a protein playfully named Sonic hedgehog is required. (This protein got its name because when a version of it is lacking in fruit fly larvae, the maggots appear rounded and bristly, like frightened hedgehogs.) Specifically, they observed that right looping occurs only if Sonic hedgehog is secreted solely on the left side of a clump of embryonic cells known as Hensen's node. (Hensen's node is the location where cells in an early chick embryo sink below other cells to create a three-dimensional embryo; a similar node occurs in mammals.) If Sonic hedgehog appears on the right side of the node instead, the developing heart loops to the left.

Sonic hedgehog is not the only player in determining the left-right asymmetry of the vertebrate heart.

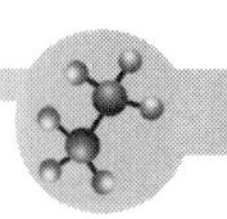

Other known proteins include Nodal and Lefty, which are secreted exclusively on the left side of an early embryo, and Activin βB, Snail and Fibroblast Growth Factor-8, which are only on the right side. When the proteins are made in their correct locations at the appropriate times during development, normal organ placement results; if the location or timing of production of any of these proteins is perturbed, abnormalities occur.

In chick embryos, for instance, the presence of Sonic hedgehog and Nodal on the left side of Hensen's node and Activin βB on the right leads to a normally asymmetric heart. Applying extra Sonic hedgehog or Nodal protein to the right side of an embryo (so that both sides of the node are now exposed to Sonic hedgehog or Nodal) can override the effects of Activin βB and confuse development: approximately half the embryos will have normal heart looping, but the heart tubes of the other half will loop in the opposite direction. The explanation for this random response seems to be that some additional factor or factors induce looping per se; in that case, Sonic hedgehog, Nodal and Activin βB influence the direction of looping. Production of Sonic hedgehog on both sides of the node leads to production of Nodal on both sides as well. Lacking clear signals as to which way to loop, each embryo "decides" on the direction of curvature randomly, resulting in 50 percent situs solitus and 50 percent situs inversus.

Interestingly, the result is the same when Sonic hedgehog or Nodal is absent from both sides. Thus,

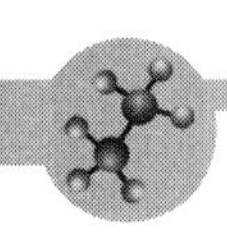

the complete absence of signals in the node or the presence of signals on both sides of the node results in random heart looping. These proteins, like all others, are made when the genes that specify their makeup are active, or switched on. It is not yet known whether people with situs inversus or isomerism have mutations in the genes for the human versions of the Sonic hedgehog and Nodal proteins, but researchers speculate that is the case.

What controls the asymmetric placement and shape of other organs? A gene recently identified by six independent research groups, including mine, seems to be part of the answer. It codes for a protein named Pitx2. Like Sonic hedgehog and Nodal, Pitx2 appears on the left side of the nascent heart and influences the direction of looping. But unlike those substances, it continues to be produced asymmetrically late into embryonic development. Moreover, it is made throughout that period on the left side of organs that are asymmetric.

Manipulating the production of Pitx2 by inserting extra copies of its gene into an embryo results in isomerism or in reverse looping of the gut and other organs as well as the heart, probably depending on the levels of the protein being made. These studies, together with experiments in which the *Pitx2* gene is inactivated, indicate that Pitx2 is one of the first factors to establish "leftness" during embryonic development. But exactly how Pitx2 and other factors result in looping of the heart tube, the coiling of the gut or the asymmetric development of the brain is still unclear.

Another open question relates to how the initial asymmetry of the body is established. What prompts the production of Sonic hedgehog, Activin βB or Lefty in the first place? One possibility is vitamin A. Over the past few years, researchers have discovered that vitamin A affects the types of cells that arise in an embryo as well as an embryo's ability to tell left from right, head from tail and back from front. They have also made great progress in understanding how vitamin A exerts these effects.

My group and others have observed, for instance, that an excess of a form of vitamin A called retinoic acid can even out the normal asymmetry of the heart in rodents and birds. It seems to do so by perturbing the production of proteins such as Nodal, Pitx2 and Lefty. Thus, it appears that the establishment of left-right asymmetry requires the exquisite regulation of vitamin A production during the early stages of embryonic development.

Other factors are certainly involved as well. Accumulating evidence suggests that specialized cell structures called cilia play a pivotal part. Cilia are whiplike structures on the outer membrane of specialized cells, such as those that line the gut; they also allow sperm to swim. Scanning electron microscopy studies have shown that all cells in the nodes of mouse embryos display a single motile cilium, located in a central position on the cell surface. The ciliated cells face the ventral (belly) side of the embryo.

In the human condition known as Kartagener's syndrome, patients have defective cilia in several cell

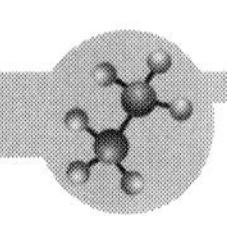

types, including sperm. These people are prone to developing respiratory infections (because they lack the cilia that normally sweep microbes out of the airways), and males are infertile. In addition, the patients tend to have situs inversus. Similarly, mice that carry a mutant form of a protein that is a component of cilia display randomized organ placement. The obvious conclusion is that the absence of functional cilia in the node causes organ positioning to be determined at random.

The Whip Factor

Astonishing findings are beginning to clarify how cilia in the node help to ensure normal organ placement. In 1998 Nobutaka Hirokawa of the University of Tokyo and his colleagues observed that mouse nodal cells, which extend their cilia into the fluid surrounding the embryo, rotate their cilia counterclockwise, in a unidirectional motion that has never been seen in other cilia. This motion, in turn, creates a flow of fluid that could sweep critical factors such as retinoic acid, Nodal and Lefty to the left side of the node. That accumulation of fluid and proteins on the left may then provide the bias required to break the initial embryonic symmetry. In other words, a feature of cellular architecture (the direction of rotation of cilia in the node) is translated into a left-right bias in embryonic development that effectively controls the way our internal organs develop.

No one understands just why the cilia rotate in a counterclockwise fashion. Presumably, though, that

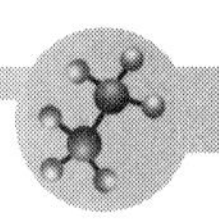

pattern arises because the molecules driving cilial motion are themselves asymmetric. Nevertheless, normal asymmetric organ placement occurs in half the individuals (people or mice) that have absolutely no cilia in their nodes. It follows that nodal cilia are not required for organ development. Rather they are needed to establish the molecular gradients that are required for the proper orientation and positioning of the organs.

When cilia are absent, the preferred flow of extraembryonic fluid fails to materialize; consequently, the left or right determinants carried by the fluid appear on both sides of the node. In such cases, organ position is established at random, presumably because of the random predominance of the appropriate chemical signals on one side of the node or the other.

The problem of left-right determination in the developing embryo has fascinated many biologists for decades, but until very recently progress was slow, in part because of the lack of molecular data. The recent discovery of genes that are active asymmetrically in the early embryo has uncovered many new clues. When some of the genes involved in a particular developmental process are known, researchers can turn them on or off in differing parts of an embryo in the laboratory to test hypotheses about the roles played by the proteins those genes encode. Although the exact nature of the initial event that establishes asymmetry in the embryo is still elusive, identification of proteins involved later on should facilitate discovery of proteins involved in other

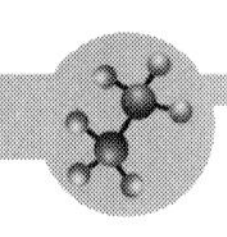

aspects of organ development. This knowledge may lead to the identification of formerly unknown mutations that predispose to specific organ malformations, which, in turn, will help in developing new systems of prenatal diagnosis.

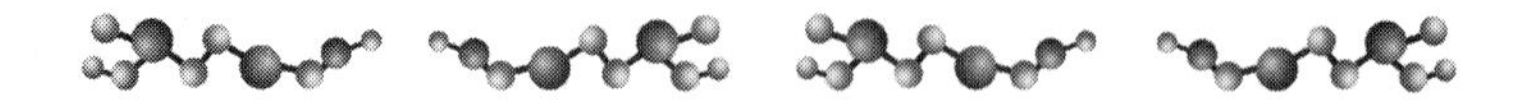

The asymmetry of expression of morphogens in one part of an embryo relative to the other part can also help to explain why, in some Siamese, or conjoined, twins, one twin has the heart, while the other does not. Siamese twins are caused by incomplete division of a fertilized egg; thus, the morphogens in one twin can affect those of the other, thereby leading to asymmetric distributions of organs between the two.

During development, cells migrate from some areas to others, induce organ formation, and then die. Some organ structures form and later disappear. For example, in a tadpole, the gills and tail that form later die and are reabsorbed. Human embryos initially have webbing between their fingers and toes, which later disappears. Many more connections between nerve cells are made than are actually used; those unused connections die out. The deaths of these cells are not due to

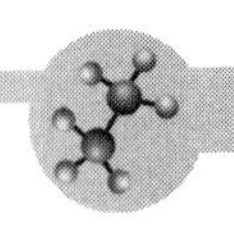

trauma or accident. Instead, they are well orchestrated and executed like commands in a computer program. This programmed cell death is called apoptosis. In the following article, Pascal Meier, Andrew Finch, and Gerard Evan discuss the molecular mechanisms involved in apoptosis. —CCF

"Apoptosis in Development"

by Pascal Meier, Andrew Finch, and Gerard Evan
***Nature*, October 2000**

Those interested in the "Big Dig," the city of Boston's heroic attempt to bury Interstate 93 beneath its pavements while maintaining a passable stab at business-as-usual above, will be well acquainted with the idea that major construction entails a substantial amount of demolition. So too in animal development: during the ontogeny of many organs, cells are over-produced only to be etched or whittled away to generate the rococo structures of functional tissues. Early distaste among biologists for the "wastefulness" of such a process has given away to the recognition that the ability to ablate cells is as essential a constructive process in animal ontogeny as are the abilities to replicate and differentiate them. After all, most animals thrive in a sea of energy and profligacy with their component cells is a small price to pay for ability to move around and propagate. It is highly unlikely that the peacock, upon encountering the peahen

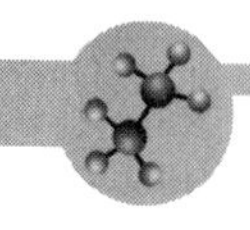

of his dreams, demurs to ponder the energetic cost of his outrageous tail.

It is now clear that physiological cell death is an essential component of animal development, important for establishment and, in vertebrates at least, maintenance of tissue architecture. A general *modus operandi* of metazoan development is the over-production of excess cells followed by an apoptotic culling during later stages of development to match the relative number of cells of different types to achieve proper organ function.[1] Thus, during animal development, numerous structures are formed that are later removed by apoptosis. This enables greater flexibility as primordial structures can be adapted for different functions at various stages of life or in different sexes. Thus, the Müllerian duct gives rise to the uterus and oviduct in females but it is not needed in males and so is consequently removed. On the other hand, the Wolffian duct is the source of male reproductive organs and is deleted in females. Organisms are like many modern computer programs, full of remnant code that was once used in an ancestral incarnation or that runs irrelevant routines that nobody needs. During development, apoptosis is frequently used to expunge such structures. For instance, early in vertebrate development, the pronephric kidney tubules arise from the nephrogenic mesenchyme. Although these pronephric tubules form functioning kidneys in fish and in amphibian larvae, they are not active in mammals and degenerate.[2] Similarly, during insect and

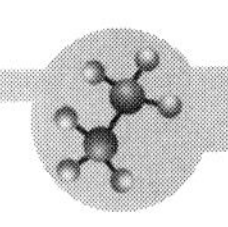

amphibian metamorphosis, apoptosis ablates cells that are no longer needed such as muscles and neurons essential for larval locomotion in insects or the amphibian tadpole tail.

Apoptosis also acts as part of a quality-control and repair mechanism that contributes to the high level of plasticity during development by compensating for many genetic or stochastic developmental errors. For example, *Drosophila* embryos with extra doses of the morphogen bicoid (*bcd*) gene show severe mispatterning in their anterior regions. Surprisingly, these embryos develop into relatively normal larvae and adults because cell death compensates for tissue overgrowth and mispatterning.[3] Cells that have been incorrectly programmed are, in effect, misplaced cells. They therefore fail to receive the appropriate trophic signals for their survival and consequently activate their innate autodestruct mechanism.

In this review we outline how each of the three great model organisms of developmental biology, the nematode *Caenorhabditis elegans*, the fruitfly *Drosophila melanogaster* and the mouse *Mus musculus* have contributed to our understanding of the role of cell death in development and homeostasis.

Lessons from Invertebrates

The first evidence for a genetic basis of apoptosis came from studies in *C. elegans* whose invariant, lineage-restricted development makes this organism particularly advantageous for the study of developmental processes.

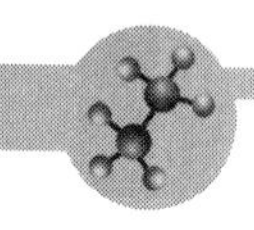

During ontogeny of the adult hermaphrodite worm, 131 of the 1,090 somatic cells die by apoptosis, leaving an adult comprised of 959 cells. Genetic screens for mutants defective in cell death identified specific genes required for regulation, execution and resolution of apoptosis, of which four, *egl-1*, *ced-3*, *ced-4* and *ced-9*, are required for each cellular demise. Loss-of-function mutations in *egl-1*, *ced-3* or *ced-4* result in survival of all 131 doomed cells, implicating these three genes in the induction of cell death. In contrast, animals lacking functional *ced-9* die early in development owing to massive ectopic cell death, whereas a gain-of-function mutation in *ced-9* blocks all 131 cell deaths, implicating *ced-9* as a suppressor of cell death.

Remarkably, this basic cell death machinery is highly conserved throughout metazoan evolution. *ced-3* encodes CED-3, a cysteine protease of an evolutionarily conserved class now dubbed "caspases" because of their predilection for cleaving at aspartyl residues. By their cleavage of critical cellular substrates, caspases act as key engines of cellular destruction in all metazoans.[4] Like most proteases, caspases are synthesized as pro-enzyme zymogens that have little or no intrinsic catalytic activity. They are activated by proteolytic cleavage either through the action of other caspases or through an autocatalytic process in which multiple procaspase molecules are brought into close proximity through formation of multiprotein "apoptosome" complexes.[5] Such complexes

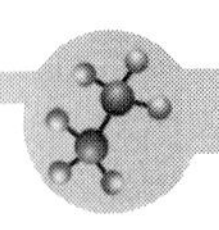

permit the low intrinsic proteolytic activity of the procaspases to trigger their own intermolecular cleavage and activation. In addition to CED-3, two other *C. elegans* caspases have been identified, CSP-1 and CSP-2 (ref. 6). However, the lack of cell death in *ced-3* mutants indicates that neither can replace CED-3 function and it is probable that both act as part of a downstream proteolytic cascade.

Caspases can be grouped into two general types based on the size of their amino-terminal prodomains. Caspases with short prodomains (type 2) are in general activated by upstream caspase cleavage and act as "effectors" that implement apoptosis by cleavage of appropriate substrates. In contrast, the extended prodomains of the so-called type 1 "initiator" caspases, of which CED-3 is an example, serve as interaction domains for assembly into "apoptosome" complexes, an assembly that is dependent on specific adaptor or scaffolding molecules and typically occurs in response to activation of some pro-apoptotic signalling pathways. In *C. elegans*, the requisite adaptor protein is encoded by *ced-4*, although its innate ability to trigger CED-3 activation is stanched by the protein product of the *ced-9* death suppressor gene. Only when CED-4 is displaced from CED-9 by the EGL-1 protein is the lethal proteolytic action of CED-3 unleashed to ablate its 131 cellular victims. Evidence indicates that EGL-1 can be regulated transcriptionally. For example, EGL-1 expression induces apoptosis in hermaphrodite-specific neurons of male

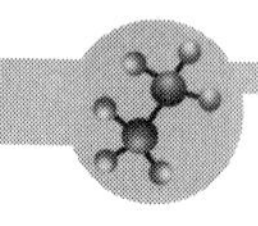

worms whereas its expression in hermaphrodites is repressed by the TRA-1A sex determination transcription factor.[7] Although EGL-1 and the CED proteins are implicated in all developmental cell deaths in *C. elegans*, not all cell deaths are regulated in the same way. For example, CES-1 and CES-2 act to regulate apoptosis in specific neurons. CES-1 is an anti-apoptotic zinc-finger transcriptional repressor of the Snail/Slug family[8] whose apoptotic target genes are unknown.[9] CES-2 is a PAS bZip protein, related to mammalian hepatocyte leukaemia factor,[10] that acts to promote apoptosis through repression of CES-1 expression.[11] Another cell type-specific example is the death of germ cells in the hermaphrodite gonad, which uses CED-3, CED-4 and CED-9 but is independent of EGL-1. This example of nematode apoptosis is also interesting because it is not pre-programmed but occurs in an adaptive way in response to DNA damage, age and environmental factors and is modulated by the Ras/mitogen-activated protein kinase (MAPK) pathway.[12, 13]

The general mechanistic interplay of the *C. elegans* cell death machinery is conserved, albeit with substantial embellishments, in other metazoans. Multiple caspases are present in both *Drosophila* and mammals, and these are in turn regulated by various homologues and analogues of the CED-4 adaptor/scaffold protein of which the evolutionarily closest known functional relatives are Apaf-1 in man[14] and dApaf-1/DARK/HAC-1 in flies.[15–17] In addition, in

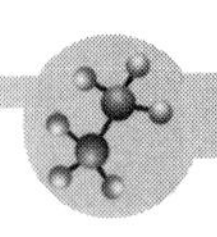

mammals at least, certain caspases are activated by recruitment into complexes induced by ligation of death receptors such as CD95 (Apo-1/Fas) and tumour necrosis factor (TNF) receptor 1. The mammalian[18] and recently identified *Drosophila*[19–22] counterparts of the CED-9 death suppressor protein are the Bcl-2 family of proteins which, in mammals (and maybe in *Drosophila*), includes both anti-apoptotic ("CED-9/Bcl-2-like") and BH3 proapoptotic ("EGL-1-like") members.[23] The remarkable conservation of molecular mechanism by which Bcl-2 proteins prevent cell death is most graphically demonstrated by the fact that human Bcl-2 is fully functional in suppressing cell deaths in developing *C. elegans*.[24]

Developmental Cell Death in *Drosophila*

It is a matter of debate whether *C. elegans* exemplifies an evolutionarily simple prototypic organism or a highly compressed and stripped down version of a more complex one. Whichever, its apoptotic machinery is clearly adapted to implementing cell death in a highly invariant manner. In contrast, development of *D. melanogaster* evidences a far greater complexity and plasticity which is mirrored in an apoptotic machinery that, through evolutionary duplication and elaboration, is significantly more complex than that of its nematode cousin. Intriguingly, this greater complexity includes a cadre of cell death regulatory proteins that so far has not been found in either nematode or mammal. Genetic analysis of *Drosophila*

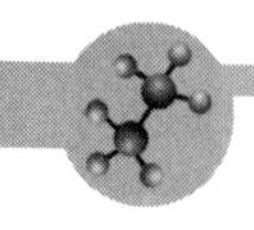

developmental cell death indicates that it is in the main determined by three pro-apoptotic proteins: reaper (RPR), GRIM and head involution defective (HID). Embryos bearing the *Df(3)H99* deletion in chromosome 3 that ablates the *rpr*, *grim* and *hid* loci show essentially no embryonic apoptosis and die towards the end of embryogenesis with a significant excess of cells.[25] Expression of two of these "death" genes, *rpr* and *grim*, is confined only to those cells destined to die: their expression presages death by some two hours. However, the *rpr* gene also triggers apoptosis in an adaptive way. rpr is induced in response to developmental malfunction, although the mechanism is unclear, and it is also a transcriptional target of the *Drosophila* p53 protein,[26–28] making its expression responsive to genotoxic stress.[28] Inducible activation of *rpr* therefore provides *Drosophila* with an adaptive mechanism for specifically ablating misplaced and genetically damaged cells.

Drosophila apoptosis is also regulated critically by survival signals provided by neighbouring cells. Thus, HID, in contrast to RPR and GRIM, is expressed in cells that both live and die but its action is regulated by the Ras/Raf/MAPK signalling pathway; this pathway promotes survival both by downregulating HID expression and by inactivating the existing HID protein through phosphorylation.[29, 30] Nonetheless, because genetic analyses indicate that *rpr*, *grim* and *hid* act cooperatively to induce apoptosis in a cell-specific manner, modulation of HID is only one of several factors

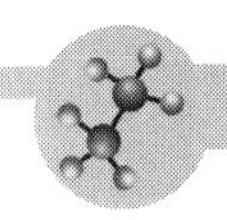

that determine cell viability in any tissue. Furthermore, at least some cell death in *Drosophila* development is independent of *rpr*, *grim* and *hid*: for example, nurse-cell apoptosis during oogenesis occurs normally in *H99* deletion flies.[31]

Whether or not RPR, GRIM and HID have any direct homologues in mammals, they nonetheless interact with components of *Drosophila* cell death machinery that are conserved, in particular the evolutionarily conserved inhibitor-of-apoptosis protein (IAP) family that in mammals are targets of the pro-apoptotic Smac/DIABLO protein.[32–34] Loss-of-function mutations in the *Drosophila* IAP DIAP-1 result in embryonic death as a result of extensive apoptosis, whereas ectopic expression of the IAPs DIAP1, DIAP2 or DETERIN suppresses cell death induced by RPR, GRIM or HID. Although some IAPs have been shown to act as direct competitive inhibitors of caspases, it seems likely that many act to bind to the large prodomains of type 1 caspases and thereby prevent their sequestration into activating apoptosome complexes. It is thought that RPR, HID and GRIM inhibit such interactions between IAPs and caspases, so promoting caspase activation and cell death.[35–38] In this way, IAPs act as "guardians" of the cell death machinery. However, in cells that are fated to die, this IAP-mediated road block has to be overcome and it seems that RPR, GRIM and HID function to do this, at least in part, by antagonizing the anti-apoptotic activity of DIAP1, thereby liberating caspases.

An ever growing number of Bcl-2 family members is emerging in both *Drosophila* and mammals, each member characterized by its ability to either induce or to suppress apoptosis.[18] Bcl-2 family proteins form homo- and heterodimers and it seems to be the net balance of protectors and killers that determines whether a cell is to live or die. It is probable that the recently identified *Drosophila* pro-apoptotic members of the Bcl-2 family such as dBORG-1/DROB-1/DEBCL/dBOK and dBORG-2/BUFFY proteins (reviewed in ref. 39) are important in determining cell survival during development. Loss of dBORG-1 function results in excess glial cells, attesting to the protein's pro-apoptotic activity,[19–22] but in other circumstances dBORG-1 seems to exert a protective effect. Unfortunately, too little is currently known of either the biochemistry or genetics of the various *Drosophila* Bcl-2 homologues to deduce their roles during fly development.

Caspases are clearly necessary for *Drosophila* cell death as their inhibition by IAPs, the baculovirus p35 caspase inhibitor or dominant-negative caspase mutants inhibits developmental apoptosis as well as apoptosis induced by overexpression of *rpr*, *hid* or *grim* (reviewed in ref. 40). Five caspases have been identified in *Drosophila*—DCP-1, drICE, DCP-2/DREDD, DRONC and DECAY—with another two predicted on the basis of genomic sequence. Both DCP-2/DREDD and DRONC have extensive prodomains suggesting that they are initiator caspases

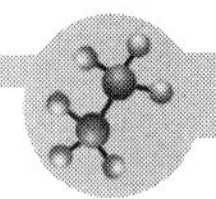

linked to specific upstream apoptotic signalling pathways. In contrast, DCP-1, drICE and DECAY all possess short prodomains and probably act as downstream "effector" caspases that are activated by upstream initiator caspases. The absence of loss-of-function mutants of most *Drosophila* caspases makes it difficult to determine their various biological role and level of redundancy. However, loss of zygotic DCP-1 leads to absence of gonads or imaginal discs (epithelial structures that will give rise to the adult insect): such animals have brittle tracheae, exhibit prominent melanotic tumours and die during larval stages. Moreover, female chimaeras bearing *dcp*$^{-1-}$ germline cells are sterile owing to a failure of nurse cells to reorganize their actin cytoskeletons, an essential process for the cytoplasmic transfer from nurse cells into the oocyte.[41]

Huge numbers of cells die during *Drosophila* embryonic and imaginal development, as well as during metamorphosis. Throughout metamorphosis, pulses of the steroid hormone 20-hydroxyecdysone (ecdysone) trigger the transition from larval to adult life and signal stage- and tissue-specific onset of apoptosis. During this transition phase the larva is profoundly reorganized to build the adult insect. Most structures that are no longer needed in the adult are deleted by apoptosis while others are newly built from imaginal precursor cells. For example, destruction of the salivary glands during metamorphosis is triggered by an ecdysone-initiated switch in gene

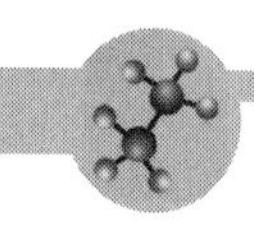

expression whereby *DIAP2* expression is down-regulated and *rpr* and *hid* expression are both induced.[42] *rpr* induction occurs by means of an ecdysone-receptor response element in the *rpr* promoter. The *rpr* promoter/enhancer element, which also contains a p53-responsive element, is extensive and contains aplethora of different response elements that serve to integrate *rpr* expression into a variety of pro-apoptotic signalling pathways.

Apoptosis in Vertebrates

In both *C. elegans* and *Drosophila*, development is restricted largely to early life and ends at birth or metamorphosis. In vertebrates, by contrast, the developmental processes of morphogenesis, remodeling and regeneration are sustained at high level in many tissues, either constitutively or in response to insult or injury, throughout their extensive life spans. The critical role of apoptosis in development is therefore evident throughout vertebrate life and, consequently, dysfunctions in apoptosis manifest themselves not only in developmental abnormalities but also in a wide variety of adult pathologies. It is these pathologies, in particular cancer and degenerative disease, that have yielded much information as to the variegated roles of apoptosis in vertebrate biology.

Vertebrate apoptotic machinery is substantially homologous to that of invertebrates, although it is more elaborate and degenerate: caspases, Bcl-2 and IAP family proteins, and survival signalling pathways

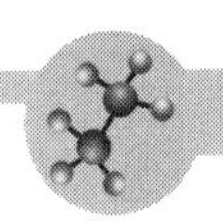

all exist in bewildering multiplicity, consistent with the more sophisticated needs for control of apoptosis in vertebrate tissues. As in invertebrates, a variety of transcriptional mechanisms is important in the regulation of developmental apoptosis in vertebrates. For example, regression of the tadpole tail during metamorphosis requires thyroid hormone-induced RNA and protein synthesis for its destruction,[43] steroid hormone receptors are critical controllers of apoptosis in many mammalian tissues including the mammary gland, the prostate, the ovary and testis, (reviewed in ref. 44) and the classical apoptotic paradigm of interdigital cell death is determined by the transcriptional readout of the transforming growth factor-β, signalling pathway (reviewed in ref. 45). Nonetheless, mammalian apoptosis is also significantly regulated, in both development and throughout life, by two non-transcriptional signalling systems whose exact counterparts are either absent or have proven elusive in the fly and worm.

First, mammals possess a family of death receptors whose ligation can directly trigger activation of specific initiator caspases through induced assembly of discrete apoptosome complexes. The archetypal members of this death-receptor family are the TNF and CD95 receptors that recruit caspase-8 via the adaptor protein FADD. Among other things, mammalian death receptors are used by cytotoxic T lymphocytes to impose an incontestable cell death programme upon target infected cells. Although no equivalent of death receptors has yet

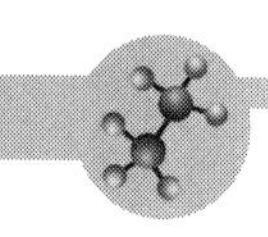

been identified in *Drosophila*, a homologue of the FADD adaptor has recently been isolated that interacts with the prodomain of the apical caspase DCP-2/DREDD (ref. 46). Furthermore, a suspicion that *Drosophila* FADD may, as in mammals, be linked to some kind of death receptor is fostered by the observation that expression of mammalian CD95 in insect cells induces apoptosis.[47]

Second, in mammals many pro-apoptotic insults seem to impact directly upon mitochondria to induce their leakiness and the release of various pro-apoptotic polypeptides. One of these is holocytochrome c, which orchestrates assembly of a complex involving Apaf-1, the closest known mammalian homologue of nematode CED-4, with caspase-9, a CARD (caspase-activating recruitment domain) initiator caspase that is then activated and triggers a downstream cascade of effector caspases.[14, 48] Another is Smac/DIABLO, which binds and antagonizes the anti-apoptotic activity of XIAP and is probably the functional analogue of *Drosophila* RPR, HID and GRIM (ref. 34). In mammals, the principal anti-apoptotic action of Bcl-2 proteins, and of the survival signalling pathways that impact on them, seems to be the stabilization of mitochondrial integrity and prevention of release of these pro-apoptotic polypeptides. This differs from the role of nematode CED-9, which acts by directly interfering with the ability of CED-4 to activate CED-3. Indeed, there is no evidence for any involvement of mitochondria in cell death in *C. elegans*. The *Drosophila*

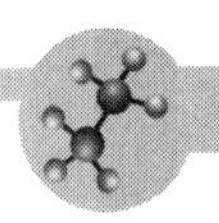

CED-4/Apaf-1 adaptor is far more similar to Apaf-1 of mammals than to CED-4 and there is clear evidence of a role for cytochrome *c* in fly apoptosis.[17, 49] Thus, so far it remains unclear whether insects and vertebrates have evolved a more complex elaboration of the apoptotic machinery that incorporates the mitochondrion or whether *C. elegans* represents a stripped down version of a more evolutionarily ancient mechanism.

For all their apparent sophistication, vertebrate tissues are constructed using the same three general principles as worms and flies—cell proliferation, differentiation and demolition. However, among other factors the size and longevity of vertebrates place peculiar demands on their apoptotic machinery. For example, although in general most vertebrate tissues may be considered no more intrinsically complex than those of invertebrates, merely larger, there are two notable exceptions. The intricacies of the vertebrate central nervous and immune systems are (with the possible exception of the sadly under-investigated cephalopods) without peer in the animal world. Both of these tissues self-assemble through an extensive iterative matching process which is governed by application of fairly simple genetically programmed rules rather than through implementation of any specific cellular map of the final organ. Such matching is an evolutionarily ancient method of construction used, for example, in the developing fly nervous system, but its extent in vertebrates is without precedent.

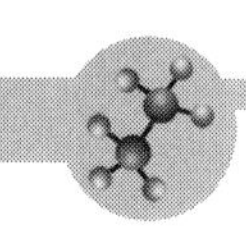

It is also an imprecise and stochastic affair that generates vast numbers of unmatched orphan cells that must be efficiently culled and removed to enable productive networks to emerge. The solution to this problem lies in configuring component cells in each tissue to commit suicide unless they establish productive connections. Thus, more than 80% of ganglion cells in the cat retina die shortly after they are born because their survival depends on the availability of limiting amounts of neurotrophic factors secreted by the target cells they innervate and for which they compete. A similar selective attrition occurs during development of the optic nerve[50] and, to various extents, in the entire central and peripheral nervous systems. In all tested cases, cell death can be largely suppressed by the excess provision of an appropriate neurotrophic survival factor. In the vertebrate immune system, cell "wastage" is even more profound: survival of the emerging lymphocyte is absolutely dependent upon the fickle assembly of a productive immune receptor that provides the trophic signal necessary to suppress apoptosis. In this way, lymphocytes bearing inoperative or self-reactive receptors are deleted from the immune repertoire.

Largely on the basis of such studies, Martin Raff first postulated that cell death is the default state of all metazoan cells which must be continuously forestalled by environmental survival signals.[51, 52] Examples of survival signals include soluble cytokines and hormones, synaptic connections, and direct

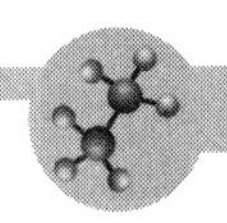

physical interactions with heterotypic cell neighbours and extracellular matrix. Different cell types require differing combinations of survival signals, which are only available within discrete somatic environments. Consequently, somatic cells are to great degree "trapped" within specialized microenvironments within the body, dying should they stray or become dislodged through injury, chance or developmental mis-programming. Perhaps the most patent examples of such somatic confinement are epithelial and endothelial cells that spontaneously commit suicide when detached from their neighbours and basal stroma because they are denied necessary integrin- and cadherin-mediated survival signals. Detachment-induced apoptosis, often termed anoikis, is an important constructive mechanism during development and triggers death in interior cells sundered from outlying basement membranes. Similar cell death also seems necessary during folding, pinching off and fusion of epithelial sheets to generate structures like lens vesicle and vertebrate neural tube. If explanted chick embryos are treated with apoptosis inhibitors the two neural folds still meet but fail to fuse to form the neural tube.[53]

The profound complexities of the nervous and immune systems make them peculiarly sensitive indicators of perturbation or dysfunction in apoptosis. Perhaps it is no surprise, therefore, that most phenotypes arising from spontaneous or induced mutations in cell death machinery are most evident in these two

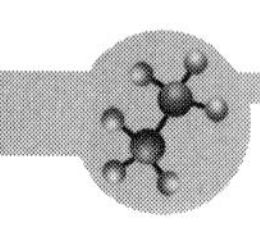

tissues. Thus, mice lacking caspase-9, caspase-3 or Apaf-1 all exhibit gross neuronal hyperproliferation and disordering,[54–58] whereas mice deficient in *bcl-x* (encoding both Bcl-xL and Bcl-xS) show a marked increase in neuronal apoptosis, leading to embryonic death.[59, 60] Pro-apoptotic signals also are important in neuronal development: for example, trophic withdrawal-induced apoptosis of spinal motor neurons is inhibited by a blocking anti-CD95 antibody.[61] Similarly, a host of immunological and haematopoietic phenotypes, some subtle and some dramatic, arise from mutations in genes that regulate or implement apoptosis.

Redundancy in Vertebrate Cell Death

In addition to their complexity, both longevity and size impose additional requirements on the vertebrate cell death machinery. Once crafted, *C. elegans* and *D. melanogaster* have highly restricted regenerative capacities (although this is not always true of other invertebrates). In contrast, apoptosis is required for tissue repair and remodeling throughout vertebrate life in order to cope with the vicissitudes of infection and injury. Unfortunately, this regenerative capacity of vertebrate tissues carries with it the risk that somatic cells will acquire mutations that confer growth independence and lead to neoplasia. Furthermore, the size and the longevity of vertebrates both conspire to increase the likelihood of such mutations occurring. It is likely that the need to have effective, overlapping and redundant mechanisms to

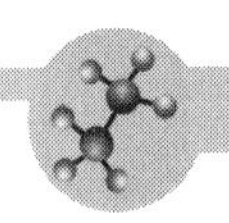

restrict the clonal autonomy of somatic cells has been one of the great imperatives of vertebrate evolution and may well have driven the remarkable redundancy in vertebrate cell death mechanisms.

The dependency of vertebrates on cell death for their greater complexity, plasticity and longevity is reflected in a far more explicit redundancy in mechanisms that regulate cell death than is apparent in invertebrates. The consequences of apoptotic dysfunction on the immune and nervous systems tend to obscure a fact of equally significant biological importance: namely, that most mutations that compromise vertebrate apoptotic machinery have surprisingly little effect on the general abilities of vertebrate tissues to develop. Although many lesions in apoptotic genes are lethal either during embryogenesis or neonatally, death typically results from focal failure in specific tissues and in no case is all embryonic apoptosis blocked. Thus, inactivation of differing caspases induces defects in specific tissues:[62, 63] for example, ablation of caspase-3 and -9 affects brain development whereas loss of caspase-8 affects heart. Likewise, the principle evident effects of loss of TNF and CD95 receptors or ligands are on the immune system.[64] Inactivation of Apaf-1, the ubiquitously expressed adaptor molecule coupling the death pathway to downstream caspases, leads to significant (but not universal) late embryonic death but defects are restricted mainly to brain and craniofacial development and to sterility in surviving males.[57, 65] Indeed, death of the interdigital webs, perhaps the

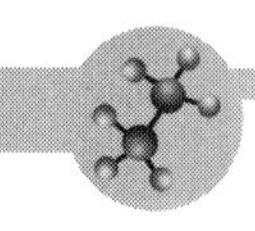

most classical paradigm of morphogenic apoptosis, proceeds unabated in Apaf-1-knockout mice. Even mice lacking cytochrome *c* survive to mid-gestation, by which time a substantial degree of apoptosis-requiring morphogenesis has already occurred.

In part, the robustness of vertebrate cell death reflects an extensive evolutionary duplication and elaboration of apoptotic machinery. At least 15 vertebrate caspases have been identified, four of which seem to be effector caspases whereas the others bear the elaborate prodomains of initiator caspases and are presumably coupled to various upstream pro-apoptotic signals. The vertebrate Bcl-2/BH3 protein family numbers some 20 current members, many of which come in "various flavours of splice." Multiple death receptors and cognate ligands have been identified and the number and diversity of signalling pathways that can regulate cell survival seem to be legion. The robustness of vertebrate cell death is probably also indicative of redundancy within the various cellular pathways that conspire to create the apoptotic process. Evidence indicates that "apoptosis" comprises at least three discrete, if intertwined, mechanisms, any one of which would alone be sufficient for cellular demise. In addition to caspase activation, most pro-apoptotic insults also trigger mitochondrial dysfunction[66] and expression of pro-phagocytic signals, neither of which necessarily depends upon caspase activity. Consistent with this, there are reported instances of inhibition of caspase activation re-routing cells into abortive or necrotic forms of cell death, but cell death nonetheless.[67–69] More

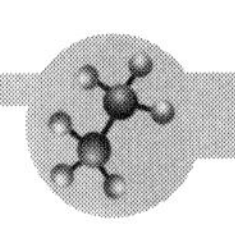

generally still, much physiological cell death in vertebrates may be triggered by nonapoptotic mechanisms. For example, in superficial epithelia such as skin and gastrointestinal tract, arguably the tissues at most risk of carcinogenic insult and neoplastic mutation, the inescapable death of progeny cells is guaranteed by a combination of irreversible post-mitotic terminal differentiation and the simple expedient of detachment and shedding. Whether such detached cells die by apoptosis or necrosis is unclear. However, the distinction is immaterial since suppression of apoptosis, for example by transgenic expression of Bcl-2 or Bcl-xL, has no observable effect on terminal differentiation and cell loss in either skin[70] or intestinal epithelium.[71]

To cope with the relentless risk of cancer, vertebrates also use cell death as an adaptive mechanism to ablate rogue or neoplastic cells. One way this is achieved is through the tight coupling of cell proliferative and apoptotic programmes such that all cells forced into a proliferative state, and therefore a potential neoplastic risk to the host, are rendered acutely sensitive to induction of apoptosis (reviewed in ref. 72). The molecular basis of this coupling seems to involve at least three independent mechanisms. First, oncoproteins like Myc or the E2F G1-progression transcription factors are potent inducers of release of cytochrome *c* from mitochondria, which can trigger activation of the Apaf-1/caspase-9 apoptotic cascade.[73] Such cytochrome *c* release is suppressed by survival signals, ensuring that activation of growth-promoting oncogenes triggers apoptosis should the affected cell or its progeny stray out of their orthodox trophic environment. Second, growth-promoting oncoproteins induce expression of p53 (ref. 74). This induces a state of extreme sensitivity to DNA damage or cellular stress, upon which the affected cell either arrests or commits suicide. Third, expression of many oncoproteins induces rapid down-regulation of cadherins, triggering a *de facto* state of anoikis unless the affected cell can expeditiously re-establish appropriate attachments.

An understanding of the mechanisms controlling and implementing apoptosis is more than a matter of mere scientific interest. Apoptosis is an essential component of most developmental abnormalities and human diseases and, in many cases, the underlying cause of the resulting pathology. Disorders associated with insufficient cell death include autoimmunity and cancer, but it has also become clear that many, if not all, viruses possess mechanisms to forestall apoptosis and provide a living host to nurture virus propagation.[75] In such cases, reinstating the blocked or defective apoptotic programme is likely to have an enormous impact on the disease. On the other hand, many other diseases including AIDS, stroke and neurodegenerative disorders such as Alzheimer's, Parkinson's and retinitis pigmentosa involve excessive apoptosis. In such instances, suppression of apoptosis may restore functionality to the affected tissue. The conservation of apoptotic machinery through evolution has provided us with a wealth of experimental systems with which to study, understand and, eventually, manipulate this fundamental biological process.

Reference

1. Jacobson, M. D., Weil, M. & Raff, M. C. Programmed cell death in animal development. *Cell* 88, 347–354 (1997).
2. Saxén, L. *Organogenesis of the Kidney* (Cambridge Univ. Press, Cambridge, 1987).
3. Namba, R., Pazdera, T. M., Cerrone, R. L. & Minden, J. S. *Drosophila* embryonic pattern repair: how embryos respond to bicoid dosage alteration. *Development* 124, 1393–1403 (1997).
4. Thornberry, N. A. & Lazebnik, Y. Caspases: enemies within. *Science* 281, 1312–1316 (1998).
5. Hengartner, M. Apoptosis. Death by crowd control. *Science* 281, 1298–1299 (1998).
6. Shaham, S. Identification of multiple *Caenorhabditis elegans* caspases and their potential roles in proteolytic cascades. *J. Biol. Chem.* 273, 35109–35117 (1998).
7. Conradt, B. & Horvitz, H. R. The TRA-1A sex determination protein of *C. elegans* regulates sexually dimorphic cell deaths by repressing the *egl-1* cell death activator gene. *Cell* 98, 317–327 (1999).
8. Inukai, T. *et al.* SLUG, a CES-1-related zinc finger transcription factor gene with antiapoptotic activity, is a downstream target of the E2A-HLF oncoprotein. *Mol. Cell* 4, 343–352 (1999).
9. Metzstein, M. M. & Horvitz, H. R. The *C. elegans* cell death specification gene *ces-1* encodes a SNAIL family zinc finger protein. *Mol. Cell* 4, 309–319 (1999).
10. Inaba, T. *et al.* Reversal of apoptosis by the leukaemia-associated E2A-HLF chimaeric transcription factor. *Nature* 382, 541–544 (1996).
11. Metzstein, M. M., Hengartner, M. O., Tsung, N., Ellis, R. E. & Horvitz, H. R. Transcriptional regulator of programmed cell death encoded by *Caenorhabditis elegans* gene *ces-2*. *Nature* 382, 545–547 (1996).
12. Gartner, A., Milstein, S., Ahmed, S., Hodgkin, J. & Hengartner, M. O. A conserved checkpoint pathway mediates DNA damage-induced apoptosis and cell cycle arrest in *C. elegans*. *Mol. Cell* 5, 435–443 (2000).
13. Gumienny, T. L., Lambie, E., Hartwieg, E., Horvitz, H. R. & Hengartner, M. O. Genetic control of programmed cell death in the *Caenorhabditis elegans* hermaphrodite germline. *Development* 126, 1011–1022 (1999).
14. Zou, H., Henzel, W. J., Liu, X., Lutschg, A. & Wang, X. Apaf-1, a human protein homologous to *C. elegans* CED-4, participates in cytochrome c-dependent activation of caspase-3. *Cell* 90, 405–413 (1997).
15. Kanuka, H. *et al.* Control of the cell death pathway by Dapaf-1, a *Drosophila* Apaf-1/CED-4-related caspase activator. *Mol. Cell* 4, 757–769 (1999).
16. Zhou, L., Song, Z., Tittel, J. & Steller, H. HAC-1, a *Drosophila* homolog of APAF-1 and CED-4 functions in developmental and radiation-induced apoptosis. *Mol. Cell* 4, 745–755 (1999).

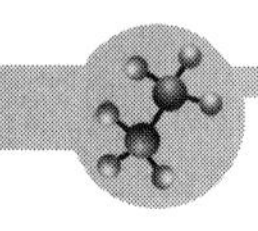

17. Rodriguez, A. *et al.* Dark is a *Drosophila* homologue of Apaf-1/CED-4 and functions in an evolutionarily conserved death pathway. *Nature Cell Biol.* 1, 272–279 (1999).
18. Antonsson, B. & Martinou, J. C. The Bcl-2 protein family. *Exp. Cell Res.* 256, 50–57 (2000).
19. Brachmann, C. B., Jassim, O. W., Wachsmuth, B. D. & Cagan, R. L. The *Drosophila* Bcl-2 family member dBorg-1 functions in the apoptotic response to UV-irradiation. *Curr. Biol.* 10, 547–550 (2000).
20. Igaki, T. *et al.* Drob-1, a *Drosophila* member of the Bcl-2/CED-9 family that promotes cell death. *Proc. Natl Acad. Sci. USA* 97, 662–667 (2000).
21. Colussi, P. A. *et al.* Debcl, a proapoptotic Bcl-2 homologue, is a component of the *Drosophila melanogaster* cell death machinery. *J. Cell Biol.* 148, 703–714 (2000).
22. Zhang, H. *et al. Drosophila* Pro-apoptotic Bcl-2/Bax homologue reveals evolutionary conservation of cell death mechanisms. *J. Biol. Chem.* 275, 27303–27306 (2000).
23. Kelekar, A. & Thompson, C. B. Bcl-2-family proteins: the role of the BH3 domain in apoptosis. *Trends Cell Biol.* 8, 324–330 (1998).
24. Vaux, D. L., Weissman, I. L. & Kim, S. K. Prevention of programmed cell-death in *Caenorhabditis elegans* by human bcl-2. *Science* 258, 1955–1957 (1992).
25. White, K. *et al.* Genetic control of programmed cell death in *Drosophila. Science* 264, 677–683 (1994).
26. Ollmann, M. *et al. Drosophila* p53 is a structural and functional homolog of the tumor suppressor p53. *Cell* 101, 91–101 (2000).
27. Jin, S. *et al.* Identification and characterization of a p53 homologue in *Drosophila melanogaster. Proc. Natl Acad. Sci. USA* 97, 7301–7306 (2000).
28. Brodsky, M. H. et al. Drosophila p53 binds a damage response element at the *reaper* locus. *Cell* 101, 103–113 (2000).
29. Bergmann, A., Agapite, J., McCall, K. & Steller, H. The *Drosophila* gene hid is a direct molecular target of Ras-dependent survival signaling. *Cell* 95, 331–341 (1998).
30. Kurada, P. & White, K. Ras promotes cell survival in *Drosophila* by down-regulating *hid* expression. *Cell* 95, 319–329 (1998).
31. Foley, K. & Cooley, L. Apoptosis in late stage *Drosophila* nurse cells does not require genes within the *H99* deficiency. *Development* 125, 1075–1082 (1998).
32. Du, C., Fang, M., Li, Y., Li, L. & Wang, X. Smac, a mitochondrial protein that promotes cytochrome c-dependent caspase activation by eliminating IAP inhibition. *Cell* 102, 33–42 (2000).
33. Verhagen, A. M. *et al.* Identification of DIABLO, a mammalian protein that promotes apoptosis by binding to and antagonizing IAP proteins. *Cell* 102, 43–53 (2000).

34. Chai, J., Du, C., Wu, J.-W., Wang, X. & Shi, Y. Structural and biochemical basis of apoptotic activation by Smac/DIABLO. *Nature* 406, 855–862 (2000).
35. Goyal, L., McCall, K., Agapite, J., Hartwieg, E. & Steller, H. Induction of apoptosis by Drosophila *reaper*, *hid* and *grim* through inhibition of IAP function. *EMBO J.* 19, 589–597 (2000).
36. Meier, P., Silke, J., Leevers, S. J. & Evan, G. I. The *Drosophila* caspase DRONC is regulated by DIAP1. *EMBO J.* 19, 598–611 (2000).
37. Song, Z. *et al.* Biochemical and genetic interactions between *Drosophila* caspases and the proapoptotic genes *rpr*, *hid*, and *grim*. *Mol. Cell Biol.* 20, 2907–2914 (2000).
38. Wang, S. L., Hawkins, C. J., Yoo, S. J., Muller, H. A. & Hay, B. A. The *Drosophila* caspase inhibitor DIAP1 is essential for cell survival and is negatively regulated by HID. *Cell* 98, 453–463 (1999).
39. Chen, P. & Abrams, J. M. *Drosophila* apoptosis and Bcl-2 genes: outliers fly in. *J. Cell Biol.* 148, 625–627 (2000).
40. Bergmann, A., Agapite, J. & Steller, H. Mechanisms and control of programmed cell death in invertebrates. *Oncogene* 17, 3215–3223 (1998).
41. McCall, K. & Steller, H. Requirement for DCP-1 caspase during *Drosophila* oogenesis. *Science* 279, 230–234 (1998).
42. Jiang, C., Baehrecke, E. H. & Thummel, C. S. Steroid regulated programmed cell death during *Drosophila* metamorphosis. *Development* 124, 4673–4683 (1997).
43. Tata, J. R. Requirement for RNA and protein synthesis for induced regression of the tadpole tail in organ culture. *Dev. Biol.* 13, 77–94 (1966).
44. Kiess, W. & Gallaher, B. Hormonal control of programmed cell death/apoptosis. *Eur. J. Endocrinol.* 138, 482–491 (1998).
45. Merino, R., Ganan, Y., Macias, D., Rodriguez-Leon, J. & Hurle, J. M. Bone morphogenetic proteins regulate interdigital cell death in the avian embryo. *Ann. NY Acad. Sci.* 887, 120–132 (1999).
46. Hu, S. & Yang, X. dFADD, a novel death domain-containing adapter protein for the *Drosophila* caspase DREDD. *J. Biol. Chem.* (in the press).
47. Kondo, T., Yokokura, T. & Nagata, S. Activation of distinct caspase-like proteases by Fas and reaper in Drosophila cells. *Proc. Natl Acad. Sci. USA* 94, 11951–11956 (1997).
48. Zou, H., Li, Y., Liu, X. & Wang, X. An APAF-1.Cytochrome c multimeric complex is a functional apoptosome that activates procaspase-9. *J. Biol. Chem.* 274, 11549–11556 (1999).
49. Varkey, J., Chen, P., Jemmerson, R. & Abrams, J. M. Altered cytochrome c display precedes apoptotic cell death in *Drosophila*. *J. Cell. Biol.* 144, 701–710 (1999).
50. Barres, B. A. & Raff, M. C. Axonal control of oligodendrocyte development. *J. Cell Biol.* 147, 1123–1128 (1999).

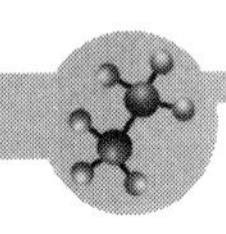

51. Raff, M. C. Social controls on cell survival and cell death. *Nature* 356, 397–400 (1992).
52. Raff, M. C. *et al.* Programmed cell death and the control of cell survival: lessons from the nervous system. *Science* 262, 695–700 (1993).
53. Weil, M., Jacobson, M. D. & Raff, M. C. Is programmed cell death required for neural tube closure? *Curr Biol* 7, 281–284 (1997).
54. Kuida, K. *et al.* Decreased apoptosis in the brain and premature lethality in CPP32-deficient mice. *Nature* 384, 368–372 (1996).
55. Kuida, K. *et al.* Reduced apoptosis and cytochrome c-mediated caspase activation in mice lacking caspase 9. *Cell* 94, 325–337 (1998).
56. Hakem, R. *et al.* Differential requirement for caspase 9 in apoptotic pathways in vivo. *Cell* 94, 339–352 (1998).
57. Yoshida, H. *et al.* Apaf1 is required for mitochondrial pathways of apoptosis and brain development. *Cell* 94, 739–750 (1998).
58. Cecconi, F., Alvarez-Bolado, G., Meyer, B. I., Roth, K. A. & Gruss, P. Apaf1 (CED-4 homolog) regulates programmed cell death in mammalian development. *Cell* 94, 727–737 (1998).
59. Middleton, G., Cox, S. W., Korsmeyer, S. & Davies, A. M. Differences in Bcl-2- and Bax-independent function in regulating apoptosis in sensory neuron populations. *Eur. J. Neurosci.* 12, 819–827 (2000).
60. Motoyama, N. *et al.* Massive cell death of immature hematopoietic cells and neurons in Bcl-x-deficient mice. *Science* 267, 1506–1510 (1995).
61. Raoul, C., Henderson, C. E. & Pettmann, B. Programmed cell death of embryonic motoneurons triggered through the Fas death receptor. *J. Cell Biol.* 147, 1049–1062 (1999).
62. Zheng, T. S., Hunot, S., Kuida, K. & Flavell, R. A. Caspase knockouts: matters of life and death. *Cell Death Differ.* 6, 1043–1053 (1999).
63. Wang, J. & Lenardo, M. J. Roles of caspases in apoptosis, development, and cytokine maturation revealed by homozygous gene deficiencies. *J. Cell Sci.* 113, 753–757 (2000).
64. Yeh, W. C., Hakem, R., Woo, M. & Mak, T. W. Gene targeting in the analysis of mammalian apoptosis and TNF receptor superfamily signaling. *Immunol. Rev.* 169, 283–302 (1999).
65. Honarpour, N. *et al.* Adult Apaf-1-deficient mice exhibit male infertility. *Dev. Biol.* 218, 248–258 (2000).
66. Green, D. R. & Reed, J. C. Mitochondria and apoptosis. *Science* 281, 1309–1312 (1998).
67. Xiang, J., Chao, D. & Korsmeyer, S. Bax-induced cell death may not require interleukin-1β-converting enzyme-like proteases. *Proc. Natl Acad. Sci. USA* 93, 14559–14563 (1996).

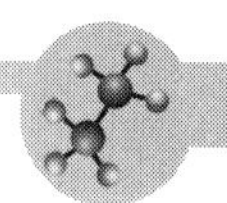

68. McCarthy, N., Whyte, M., Gilbert, C. & Evan, G. Inhibition of Ced-3/ICE-related proteases does not prevent cell death induced by oncogenes, DNA damage, or the Bcl-2 homologue Bak. *J. Cell Biol.* 136, 215–227 (1997).
69. Chautan, M., Chazal, G., Cecconi, F., Gruss, P. & Golstein, P. Interdigital cell death can occur through a necrotic and caspase-independent pathway. *Curr. Biol.* 9, 967–970 (1999).
70. Pena, J. C., Fuchs, E. & Thompson, C. B. Bcl-x expression influences keratinocyte cell survival but not terminal differentiation. *Cell Growth Differ.* 8, 619–629 (1997).
71. Merritt, A. J. *et al.* Differential expression of *bcl-2* in intestinal epithelia. Correlation with attenuation of apoptosis in colonic crypts and the incidence of colonic neoplasia. *J. Cell Sci.* 108, 2261–2271 (1995).
72. Evan, G. & Littlewood, T. A matter of life and cell death. *Science* 281, 1317–1322 (1998).
73. Juin, P., Hueber, A. O., Littlewood, T. & Evan, G. c-Myc-induced sensitization to apoptosis is mediated through cytochrome c release. *Genes Dev.* 13, 1367–1381 (1999).
74. Sherr, C. J. & Weber, J. D. The ARF/p53 pathway. *Curr. Opin. Genet. Dev.* 10, 94–99 (2000).
75. Roulston, A., Marcellus, R. C. & Branton, P. E. Viruses and apoptosis. *Annu. Rev. Microbiol.* 53, 577–628 (1999).

3 Natural Modes of Differentiation

The ability to regrow body parts or an entire organism from a single portion is called regeneration. For example, if you cut the arm off a starfish, a new one will grow back. Likewise, if you cut a flatworm (planaria) in half, a new half will grow from the severed half to make a whole animal. A popular student experiment is to cut the head of a planaria in half without severing the entire animal, and it will regrow two heads! In humans, the liver is the only organ capable of regeneration. Besides repair, some organisms use regeneration as a means of asexual reproduction.

In regeneration, specialized cells in an organ or organism undergo renewed growth, cell division, and differentiation. In some organs, non-differentiated stem cells exist that carry out this process. In the following article, Seh-Hoon Oh, Heather M. Hatch, and Bryon Petersen discuss how regeneration in the liver occurs by recruiting oval stem cells. —CCF

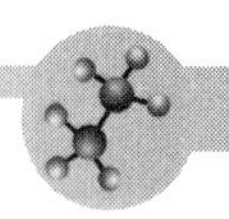

"Hepatic Oval 'Stem' Cell in Liver Regeneration"

by Seh-Hoon Oh, Heather M. Hatch, and Bryon E. Petersen

Seminars in Cell & Developmental Biology, 2002

The adult mammalian liver has been shown to have a high regenerative capacity. The liver is composed mainly of two epithelial cell types, which are the hepatocytes and bile ductular cells. These cells have a remarkable capacity to meet replacement demands incurred during normal or mild cellular loss. In rodents, loss of liver mass, up to two-thirds, due to surgical (partial hepatectomy (PHx)) or chemical means induces the normal regenerative process. The regenerative process repairs the liver and brings it back to its original mass within 10 days post hepatic injury.[1] However, when 2-acetylamino fluorene (2-AAF) is given to the animal continuously, the process is slowed due to the lack of epithelial cell division. The liver utilizes a new source of cells to help repair the liver, these are called "oval cells." Oval cells are recognized as playing an important role in the etiology of hepatic growth and development.[2–5] Furthermore, hepatic oval cells are capable of differentiation into several lineages, which include bile ductal epithelia, hepatocytes, intestinal epithelial, and possibly exocrine pancreas.[6–9] A recent *in vitro* study has shown that oval cells can be coaxed into insulin producing pancreatic cells and could lower the glucose levels of a diabetic mouse.[10] While it is has been shown that hepatic oval cells play a part in the regeneration of

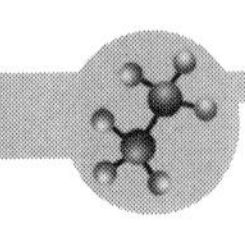

a severely damaged liver, the mechanisms by which they accomplish this are still as yet poorly understood.

Stem cells are defined as cells that are undifferentiated, capable of self-renewal, with potential to differentiation into multiple lineages and having the flexibility to use all of these options. Out of these characteristics, it would appear that differentiation plasticity is most important. Through recent progress that has been made in adult stem cell research, we now can identify various stem/progenitor cells derived from adult tissue. Hematopoietic stem cells, neural stem cells,[11] vascular endothelial progenitor cells,[12] and hepatic oval cells[13] are just a few among this group.

During liver regeneration induced by partial hepatectomy in adult rats and mice, DNA synthesis is initiated in the remaining liver cells, followed by division of the cells and the subsequent regeneration of the liver. The factors related to liver regeneration have been a topic of intense scrutiny for several decades. Regeneration of the liver is tightly controlled by growth factor(s), cytokine(s) and other factors. Hepatocyte growth factor (HGF), was originally identified and cloned as a potent mitogen for hepatocytes.[14, 15] It also shows mitogenic, motogenic and morphogenic activities for a wide variety of cells that express the HGF receptor c-Met, which is a transmembrane protein that possesses an intracellular tyrosine kinase domain.[16, 17] Moreover, HGF has been shown to play an essential role in the regeneration of the liver as well as in liver development.[1, 18, 19]

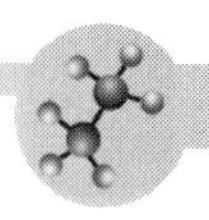

Hepatic Oval "Stem" Cell

Hepatic oval "stem" cells are recognized as participating in liver regeneration under certain conditions, as well as being implicated in hepatic carcinogenesis.[2-5] When liver damage is severe enough that large numbers of hepatocytes are lost and/or their proliferation is prevented by exposure to hepatotoxins or carcinogens, liver oval cells appear in the periportal regions of the liver. Oval cells are small cells with a large nucleus to cytoplasm ratio, in which the nucleus also has a distinctive ovoid shape, hence the name "oval."[2] They are thought to have the ability to proliferate-clonogenically as well as possess a bipotential capacity, which allows them to differentiate into both hepatocytes and bile ductular cells.

To induce oval cell proliferation in the rat liver, the combination of 2-AAF treatment with either a two-thirds partial hepatectomy (PHx) or LD50 dose of carbon tetrachloride (CCl_4) is often used.[20] Oval cell activation in this model uses the continuous administration of a low dose of 2-AAF to suppress the proliferation of hepatocytes. Approximately 7 days post-2-AAF exposure, the animals are subjected to a severe hepatic injury either through PHx or an LD_{50} dose of CCl_4. The lack of response by the hepatocytes to growth signals results in the activation and rapid proliferation of oval cells.[13] The oval cells initially appear near bile ductules, followed by migration into the hepatic parenchyma. The origins of oval cells have been questioned, although most researchers tend to believe

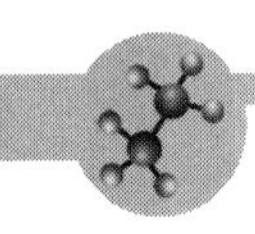

that the cells reside in the canals of Hering, which is a transitional zone between the periportal hepatocytes and the biliary cells lining the smallest terminal bile ducts, recent studies clearly indicate that at least a portion of the oval cells can be derived from bone marrow cells.[20–24] In addition, Oh *et al.* reported that bone marrow cells could be differentiated into a hepatic lineage *in vitro*.[25] Detailed tracking experiments revealed that oval cells first differentiate into basophilic small hepatocytes and then into mature adult hepatocytes.[2, 26]

In the rat liver, it has been shown that oval cells are capable of differentiating into two separate lineage *in vivo*, hepatocytes[3, 27] and intestinal type epithelium.[28] Furthermore, oval cells in culture may be induced to differentiate into both hepatocyte-like and biliary-like cell lines as well as pancreatic-like cells.[10, 29, 30] Markers commonly used to assess differentiation and to trace lineages of oval cells include expressed antigenic markers for hepatocytes, bile ducts and oval cells (BSD7, OC2, OC3, OV-1, and OV-6), intermediate filaments, extracellular matrix proteins (CK8, 18, 19), enzymes and secreted proteins (alpha-fetoprotein, gamma-glutamyl transferase).[31, 32] In addition, it has been demonstrated that oval cells also express Thy-1, CD-34 and c-kit as well as Flt-3, all of which are all known to be hematopoietic stem cell markers.[33–36] Recently it has been shown that through anti-Thy-1 FITC conjugated antibody labeling and sorting via flow cytometry, it is possible to obtain a 95–97% pure population of Thy-1$^+$ oval cells. To date, this is the highest reported purity for isolated oval cells. In addition to Thy-1, these cells have

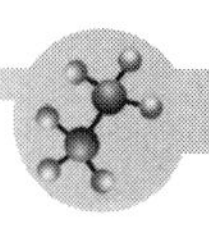

also been shown to express the traditional oval cell makers of AFP, CK-19, GGT, OC2, and OV-6. It is also important to point out that these cells did not express desmin, which is a stellate cell marker.[36]

Growth Factors

During normal hepatic regeneration as well as during renewal from the stem cell compartment, several growth factors appear to effect the proliferation and differentiation of hepatic oval cells.[37, 38] The question therefore arises as to whether the same growth factors known to be involved in normal hepatic regeneration are also involved in regeneration via the stem cell compartment.

There are three "primary" growth factors associated with normal hepatic regeneration: hepatocyte growth factor (HGF), epidermal growthfactor (EGF), and transforming growth factor-α (TGF-α). These three factors have been shown to be strong mitogens for primary hepatocytes and to stimulate DNA synthesis in chemically defined medium as well as being involved in liver regeneration.[1, 39] HGF has also been shown as being able to induced differentiation of bone marrow cells into a hepatic lineage *in vitro*.[25] Oval cells express receptors for all of these growth factors,[38, 41, 42] providing a molecular pathway by which stellate cells may influence the growth and development of oval cells. In addition, transforming growth factor-beta 1 (TGF-β1) is also expressed during hepatic regeneration, and it has been proposed that TGF-β1 may provide at least a portion of the negative growth signal

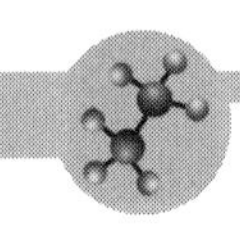

following compensatory hyperplasia that occurs following the loss of liver mass.[40]

Levels of HGF increase remarkably in the plasma and livers of mice and rats exposed to CCl_4 or subjected to PHx.[1, 18, 19, 43] An elevation of the mRNA levels of HGF in the spleen and lung are also seen after PHx.[44–48] Furthermore, in regeneration of the liver after PHx, the increase of HGF mRNA in mesenchymal cells occurs in parallel with a rise in TGF-α and acidic fibroblast growthfactor production.[49, 50] In fact, injection of HGF, TGF-α, or EGF into rats has been shown to induce DNA synthesis in the *in vivo* hepatocyte directly.[39, 51] This has also been shown to be the case in circumstances which include pretreatment of the animal with manipulators such as nutritional alterations[52, 53] and collagenase perfusion.[54] Thus, these growth factors seem to play an important role in compensatory liver regeneration.

The Role of the Oval Cell in Hepatic Regeneration

Oval cell activation is the result following a well orchestrated set of events that occur during specific times during the process of oval cell proliferation and differentiation in the liver. In the 2-AAF/PHx model, the stem cell factor/c-kit (SCF/c-kit) ligand/receptor system and α-fetoprotein expression are elevated in the earliest stages of liver oval cell activation and proliferation.[34, 55] These results suggest that the activation and expansion of hepatic oval cells is controlled during stem cell–driven liver regeneration. Recent studies show that a chemokine known as

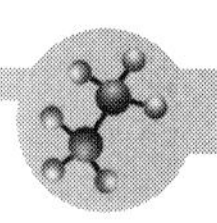

stromal derived factor-1 alpha (SDF-1) is produced in the liver and released into the damaged tissue.[56] CXCR4-positive bone marrow stem cells could then be recruited to the site of injury via chemotactic gradient. As the progenitor cells enter the liver, they come into contact with another chemokine, SCF, which has also been shown to facilitate the recruitment of stem cells.[57, 58] Concurrently, there is an expansion in the number of stellate cells in the periportal regions of the liver.[59] This results in an increased production of the growth factors such as HGF and TGF-α[42, 60] HGF has been shown to act as a strong promoter of differentiation toward the hepatic lineages.[1] It should also be noted that oval cells and bone marrow stem cells express the HGF receptor c-Met.[7, 17, 24, 31, 61] There is an increase in soluble fibronectin, an extracellular matrix (ECM) molecule produced by the stellate cells, which provides another avenue stellate cells can interact in stem cell engraftment.[60] Also, with the initiation of DNA synthesis in OV-6-positive and desmin-positive cells in portal tracts, expression of TGF-α, HGF, aFGF, and TGF-β1 are observed. These growth factors continue to be expressed at high levels throughout the period of expansion and differentiation of the oval cell population.[38, 42, 60] It is interesting to note that in a model of active proliferation, high levels of TGF-β1 are found. To address this phenomenon, for bone marrow derived stem cells to differentiate down the hepatic lineage, the differentiation down the hematopoietic lineage must be blocked. TGF-β1 and TNF have been shown to suppress the differentiation of progenitor HSCs into megakaryocytes

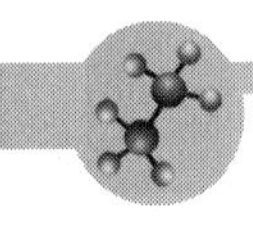

and down the myeloid lineage.[62, 63] Thus, the requisite contributors to the processes of homing, engrafting and differentiation would be in the correct alignment to signal the oval cell for its final role in the liver. These factors, as well as other as of yet undetermined factors, could be the signals needed to cause oval cells to differentiate down the hepatic lineage. Therefore, future experiments of hepatic oval cell research will be aimed at elucidating the mechanism of activation and differentiation of the cells as well as to clarify the role of the hepatic/hematopoietic interaction of oval cell aided liver regeneration. This may, in turn, lead to a clinical relevance with an application in the treatment of patients through stem cells therapies.

References

1. Michalopoulos GK, DeFrances MC (1997) Liver regeneration. Science 276:60–66
2. Farber E (1956) Similarities in the sequence of early histologic changes induced in the liver of the rat by ethionine, 2-acetylaminofluorene, and 3′-methyl-4-dimethyaminoazbenzene. Cancer Res 16:142–151
3. Evarts RP, Nagy R, Marsden E, Thorgeirsson SS (1987) A precursor product relationship exists between oval cells and hepatocytes in rat liver. Carcinogenesis 8:1737–1740
4. Farber E (1991) Hepatocyte proliferation in stepwise development of experimental liver cell cancer. Digest Dis Sci 36:973–978
5. Fausto N, Lemire JM, Shiojiri N (1992) Oval cells in liver carcinogenesis: cell lineages in hepatic development and identification of stem cells in normal liver, in The Role of Cell Types in Hepatocarcinogenesis, Chapter 5 (Sirica AE, ed.) p. 89. CRC Press, Boca Raton, FL
6. Thorgeirsson SS (1993) Hepatic stem cells. Am J Pathol 142:1331–1333
7. Fausto N (1990) Oval cells and liver carcinogenesis: an analysis of cell lineages in hepatic tumors using oncogene transfection techniques. Prog Clin Biol Res 331:325–334
8. Sell S (1990) Is there a liver stem cell? Cancer Res 50:3811–3815

9. Sigal SH, Brill S, Reid LM (1992) The liver as a stem cell and lineage system. Am J Physiol 263:G139–G148
10. Yang LJ, Li SW, Hatch HM, Ahrens K, Cornelius JG, Petersen BE, Peck AB (2002) *In vitro* trans-differentiation of adult rat hepatic stem cells into endocrine hormone-producing cells. Proc Natl Acad Sci USA 99(12):8078–8083
11. Steindler DA, Pincus DW (2002) Stem cells and neuropoiesis in the adult human brain. Lancet 359:1047–1054
12. Scott EW, Simon MC, Anastasi J, Singh H (1994) Requirement of transcription factor PU.1 in the development of multiple hematopoietic lineages. Science 265:1573–1577
13. Petersen BE, Zajac VF, Michalopoulos GK (1998) Hepatic oval cell activation in presence to injury following chemically induced periportal or pericentral damage in rats. Hepatology 27:1030–1038
14. Nakamura T, Nawa K, Ichihara A (1984) Purification and characterization of hepatocyte growth factor from serum of hepatectomized rats. Biochem Biophys Res Commun 122: 1450–1459
15. Russell WE, McGowan JA, Bucher NLR (1984) Partial characterization of a hepatocyte growth factor from rat platelets. J Cell Physiol 119:184–192
16. Rubin JS, Bottaro DP, Aaronson SA (1993) Hepatocyte growth factor/scatter factor and its receptor, the c-met proto-oncogene product. Biochim Biophys Acta 1155:357–371
17. Zarnegar R, Michalopoulos GK (1995) The many faces of hepatocyte growth factor: from hepatopoiesis to hematopoisis. J Cell Biol 129:1177–1180
18. Boros P, Miller CM (1995) Hepatocyte growth factor: a multifunctional cytokine. Lancet 345:293–295
19. Schmidt C, Bladt F, Goedecke S, Brinkmann V, Zschiesche W, Sharpe M, Gherardi E, Birchmeier C (1995) Scatter factor/ hepatocyte growth factor is essential for liver development. Nature 373:699–702
20. Petersen BE, Bowen WC, Patrene KD, Mars WM, Sulivan AK, Murase N, Boggs SS, Greenberger JS, Goff JP (1999) Bone marrow as a potential source of hepatic oval cells. Science 284:1168–1179
21. Theise ND, Badve S, Saxena R, Henegariu O, Sell S, Crawford JM, Krause DS (2000) Derivation of hepatocytes from bone marrow cells in mice after radiation-induced myeloablation. Hepatology 31:235–240
22. Theise ND, Nimmakayalu M, Gardner R, Illei PB, Morhan G, Teperman L, Henegariu O, Krause DS (2000) Liver from bone marrow in humans. Hepatology 32:11–16
23. Alison MR, Poulsom R, Jeffery R, Dhillon AP, Quaglia A, Jacob J, Novelli M, Prentice G, Williamson J, Wright NA (2000) Hepatocytes from non-hepatic adult stem cells. Nature 406:257

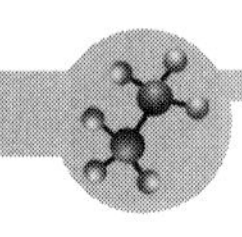

24. Lagasse E, Connors H, Al-Dhalimy M, Reitsma M, Dohse M, Osborne L,Wang X, Finegold M,Weissman IL, Grompe M (2000) Purified hematopoietic stem cells can differentiate into hepatocytes *in vivo*. Nat Med 6:1212–1213
25. Oh S-H, Miyazaki M, Kouchi H, Inoue Y, Sakaguchi M, Tsuji T, Shima N, Higashio K, Namba M (2000) Hepatocyte growth factor induces differentiation of adult rat bone marrow cells into a hepatocyte lineage *in vitro*. Biochem Biophys Res Commun 279:500–504
26. Golding M, Sarraf CE, Lalani EN, Anilkumar TV, Edwards RJ, Nagy P, Thorgeirsson SS, Alison MR (1995) Oval cell differentiation into hepatocytes in the acetylamino-fluorene-treated regenerating rat liver. Hepatology 22:1243–1253
27. Lemire JM, Shiojiri N, Fausto N (1991) Oval cell proliferation and the origin of small hepatocytes in liver injury induced by d-galactosamine. Am J Pathol 139:535–552
28. Tatematsu M, Kaku T, Medline A, Farber E (1985) Intestinal metaplasia as a common option of oval cells in relation to cholangiofibrosis in the livers of rats exposed to 2-acetylaminofluorene. Lab Invest 52:354–362
29. Hayner NT, Braun L, Yaswen P, Brooks M, Fausto N (1984) Isozyme profiles of oval cells, parenchymal cells, and biliary cells isolated by centrifugal elutriation from normal and prereoplastic liver. Cancer Res 44:332–338
30. Germain L, Goyette R, Marceau N (1985) Differential cytokeratin and alpha-fetoprotein expression in morphologically distinct epithelial cells emerging at the early stages of rat hepatocarcinogenesis. Cancer Res 45:673–681
31. Germain L, Noel H, Gourdeau H, Marceau N (1988) Promotion of growth and differentiation of rat ductular oval cells in primary culture. Cancer Res 48:368–378
32. Thorgeirsson SS (1998) Hepatic stem cells in liver regeneration. FASEB J 10:1249–1256
33. Omori N, Omori M, Evarts RP, Teramoto T, Miller MJ, Hoang TN, Thorgeirsson SS (1997) Partial cloning of rat CD34 cDNA and expression during stem cell-dependent liver regeneration in the adult rat. Hepatology 26:720–727
34. Omori N, Omori M, Evarts RP, Teramoto T, Thorgeirsson SS (1997) Coexpression of flt-3 ligand/flt-3 and SCF/c-kit signal transduction system in bile-duct-ligated SI and W mice. Am J Pathol 150:1179–1187
35. Fujio K, Evarts RP, Hu Z, Marsden ER, Thorgeirsson SS (1994) Expression of stem cell factor and its receptor, c-kit, during liver regeneration from putative stem cells in the adult rat. Lab Invest 70:511–516
36. Petersen BE, Goff JP, Greenberger JC, Michalopoulos GK (1998) Hepatic oval cells express the hematopoietic stem cell marker Thy-1 in the rat. Hepatology 27:433–445

37. Evarts RP, Nakatsukasa H, Marsden ER, Hu Z, Thorgeirsson SS (1992) Expression of transforming growthfactor-α in regeneration liver and during hepatic differentiation. Mol Carcinogen 5:25–31
38. Evarts RP, Hu Z, Fujio K, Marsden ER, Thorgeirsson SS (1993) Activation of hepatic stem cell compartment in the rat: role of transforming growth factor-α, hepatocyte growth factor, and acidic fibroblast growth factor in early proliferation. Cell Growth Differ 4:555–561
39. Marsden ER, Hu Z, Fujio K, Nakatsukasa H, Thorgeirsson SS (1992) Expression of acidic fibroblast growth factor in regeneration liver and during hepatic differentiation. Lab Invest 67:427– 433
40. Michalopoulos GK (1990) Liver regeneration: molecular mechanisms of growth control. FASEB J 4:176–187
41. Mead JE, Fausto N (1989) Transforming growth factor-α may be a phygiological regulator of liver regeneration by means of autocrine mechanisms. Proc Natl Acad Sci USA 86:1558–1562
42. Lenzi R, Liu MH, Tarsetti F, Slott PA, Alpini G, Zhai WR *et al.* Histogenesis of bile duct-like cells proliferating during ethionine hepatocarcinogenesis. Evidence for a biliary epithelial nature of oval cells. Lab Invest 66:390–402
43. Hu Z, Evarts RP, Fujio K, Marsden ER, Thorgeirsson SS (1993) Expression of hepatocyte growth factors and c-met genes during hepatic differentiation and liver development in the rat. Am J Pathol 142:1823–1830
44. Lindroos PM, Zarnegar R, Michalopoulos GK (1991) Hepatocyte growth factor (hepatopoietin A) rapidly increases in plasma before DNA synthesis and liver regeneration stimulated by partial hepatectomy and carbon tetrachloride administration. Hepatology 13:743–749
45. Nakamura T, Nishizawa T, Hagiya M, Seki T, Shimonishi M, Sugimura A, Tashiro K, Shimizu S (1989) Molecular cloning and expression of human hepatocyte growth factor. Nature 342: 440–443
46. Okajima A, Miyazawa K, Kitamura N (1990) Primary structure of rat hepatocyte growth factor and induction of its mRNA during liver regeneration following hepatic injury. Eur J Biochem 193:375–381
47. Selden C, Jones M, Wade D, Hodgson H (1990) Hepatotropin mRNA expression in human fetal liver development and liver regeneration. FEBS Lett 270:81–84
48. Zarnegar R, DeFrances MC, Kost DP, Lindroos P, Michalopoulos GK (1991) Expression of hepatocyte growth factor mRNA in regenerating rat liver after partial hepatectomy. Biochem Biophys Res Commun 177:559–565
49. Yanagita K, Nagaike M, Ishibashi H, Niho Y, Matsumoto K, Nakamura T (1992) Lung may have an endocrine function producing hepatocyte growth factor in response to injury of distal organs. Biochem Biophys Res Commun 182:802–809

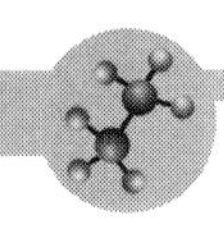

50. Kan M, Huang JS, Mansson PE, Yasumiysu H, Carr B, Mckeehan WL (1989) Heparin-binding growth factor type I (acidic fibroblast growth factor): a potential biphasic autocrine and paracrine regulator of hepatocyte regeneration. Proc Natl Acad Sci USA 86:7432–7436
51. Bucher NL, Patel U, Cohen S (1977) Hormonal factors concerned with liver regeneration. Ciba Found Symp 55:95–107
52. Ishiki Y, Ohnishi H, Muto Y, Matsumoto K, Nakamura T (1992) Direct evidence that hepatocyte growth factor is a hepatotrophic factor for liver regeneration and has a potent antihepatitis effect *in vivo*. Hepatology 16:1227–1235
53. Webber EM, Godowski PJ, Fausto N (1994) *In vivo* response of hepatocytes to growth factors requires an initial priming stimulus. Hepatology 20:489–497
54. Roos F, Ryan AM, Chamow SM, Bennett GL, Schwall RH (1995) Induction of liver growth in normal mice by infusion of hepatocyte growth factor/scatter factor. Am J Physiol 268:G380–G386
55. Liu ML, Mars WM, Zarneger R, Michalopoulos GK (1994) Collagenase pretreatment and the mitogenic effects of hepatocyte growth factor and transforming growth factor-α in adult rat liver. Hepatology 19:1521–1527
56. Hatch HM, Jorgensen ML, Stolz DB, Petersen BE (2001) SDF-1 as a potential homing protein for bone marrow derived liver oval cells. FASEB J 15(5) Part II:A1084, Abstract #827.5
57. Fujio K, Evarts RP, Hu Z, Marsden ER, Thorgeirsson SS (1994) Expression of stem cell factor and its receptor, c-kit, during liver regeneration from putative stem cells in adult rat. Lab Invest 70:511–516
58. Broxmeyer HE, Kim CH (1999) Regulation of hematopoiesis in a sea of chemokine family members with a plethora of redundant activities. Exp Hematol 27:1113
59. Yin L, LynchD, Sell S (1999) Participation of different cell types in the restitutive response of the rat liver to periportal injury induced by allyl alcohol. J Hepatol 31:497–507
60. Grisham JW, Thorgeirsson SS (1997) Liver stem cells, in Stem Cell (Potten CS, ed.) p. 233. Academic Press, San Diego, CA
61. Goff JP, Shields DS, Petersen BE, Zajac VF, Michaelopoulos GK, Greenberger JS (1996) Synergistic effects of hepatocyte growth factor on human cord blood $CD34^+$ progenitor cells are the result of c-met receptor expression. Stem Cells 14:592–602
62. Han ZC, Lu M, Li J, Defard M, Boval B, Schlegel N, Caen JP (1997) Platelet factor 4 and other CXC chemokines support the survival of normal hematopoietic cells and reduce the chemosensitivity of cells to cytotoxic agents. Blood 89:2328– 2335
63. Pierelli L, Marone M, Bonanno G, Mozzetti S, Rutella S, Morosetti R, Rumi C, Mancuso S, Leone G, Scambia G (2000) Modulation of bcl-2 and p27 in

human primitive proliferating hematopoietic progenitors by autocrine TGF-beta 1 is a cell cycle-independent effect and infuences their hematopoietic potential. Blood 95:3001–3009

Cancer has been traditionally thought of as a disease of abnormal control of cell division. Most noncancer cells usually go through several phases of the cell division cycle (up to thirty) and then stop. Mutations in the genes that control cell division can lead to tumor formation. These mutations can be brought about by environmental situations such as chemicals, toxins, and ultraviolet light or by viruses that can activate cancer-causing genes such as oncogenes and proto-oncogenes in the normal cell. However, in recent years, findings in developmental biology have influenced the thinking of some cancer researchers, encouraging them to look at cancer in new ways. The discovery of morphogens and how morphogen gradients produce patterns in tissue have led to the idea that cancer may be a disease of abnormal differentiation. In the following article, John D. Potter hypothesizes how morphogen-like molecules may be responsible for normal tissue structures and how disruption of these molecules might lead to tumors. —CCF

From "Morphostats: A Missing Concept in Cancer Biology"

by John D. Potter

***Cancer Epidemiology, Biomarkers & Prevention*, March 2001**

. . . The earliest observable change in the progression toward cancer is disruption of tissue architecture. Therefore, cancer is not just a function of cells (see Ref.3) and, despite opinions to the contrary,[4] is not specifically a function of proliferation; sometimes there are excess numbers of cells, sometimes there is cell loss, and sometimes there is metaplasia, but there is always disruption of architecture.

Bissell *et al.*[5] have shown that cross-talk is central to the cancer process. The hypothesis advanced here is that a key element is the loss of a specific kind of instructional cross-talk, namely, that which is responsible for maintaining tissue architecture.

Models of cancer are focused almost exclusively on cells and molecules.[6] I have previously suggested, nonetheless, that it is possible to construct models of cancer at the level of population, organism, organ, cell, and molecule [7] and that these are not all reducible to the molecular level. Certainly, others have argued that cancers themselves should be regarded as complex tissues in their own right.[3, 6, 8–12] Here I argue that cancer is importantly characterized not only by cell and molecular dysfunction but by loss of normal tissue microarchitecture and that this loss is a (perhaps *the*) fundamental step in carcinogenesis.

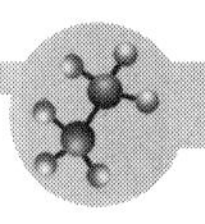

The Generation of Form in Tissues

Tissues are complex mixtures of cells. They are constructed via an intricate set of processes that involve, even in simple organisms, multiple signaling pathways, stepwise and integrated differentiation, proliferation, apoptosis, cell migration, etc. Most of these same processes persist in replicating postembryonic (and, in adults, largely epithelial) tissues. Blood cells also undergo controlled replication throughout life but are not organized into tissues. Cells of the immune system exist in both states.

In reference to the process of tissue construction and morphogenesis, Lawrence and Struhl[13] describe the class of fundamental organizers of tissue morphology—morphogens—as follows: "a morphogen emanates from a localized source and diffuses away to make a concentration gradient." They further note that the complete morphogen does more than just turn genes "on" or "off" at different concentrations; it orchestrates cellular behavior coherently so that its distribution prefigures the pattern. Hence, if the distribution changes, "even details of the pattern change in a predictable and coordinated way." Perhaps the most complete understanding of the process of constructing an organism has been generated for *Drosophila*.[14–19]

In the context of developmental biology, Ettinger and Doljanski[20] define pattern formation, morphogenesis, and differentiation. However, they do not provide a definition of the process that is involved when, in adult tissue, microarchitectural form is maintained. The

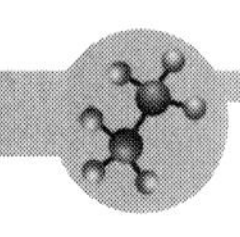

process in adult tissue includes a limited form of morphogenesis as well as differentiation, and thus, many of the ideas that they explicate are appropriate in the context of adult tissues as well. For instance, they stress the importance of the extracellular matrix (ECM) in molding the architecture of tissue and note that the basement membrane forms the scaffold of epithelial tissue.[20, 21] They note that elimination "of the basement membrane results in loss of the tissue-specific structure."[20] They also note the role of the ECM in determining the "condensations" that are characteristic of mesenchyme that is organizing itself into an instructive unit, here in the setting of limb morphogenesis, but such condensations are a more general phenomenon in the initiation of morphogenesis of many organs.[22] They note that although the mechanism underlying the condensation process remains unknown, there are a variety of possibilities, but it seems likely that important signaling pathways including those involving transforming growth factor (TGF-β) and basic fibroblast growth factor and cell adhesion molecules such as cadherins are involved[20] . . .

Adult epithelia face the same issues as embryonic tissues: how to maintain an overall form while constituent elements (cells) proliferate, move, differentiate, and die. However, adult epithelia face additional issues: the need to restore integrity after injury and the need to avoid chronic damage. Despite these challenges, most adult tissues do maintain a consistent microarchitecture over many decades. It is reasonable to ask whether the same kinds of controls and signals that build the tissues

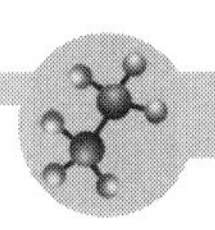

in the embryo have counterparts in the maintenance of the adult architecture. Hodges[33] notes that the stroma possesses "certain qualities that generate specific genetic expression and regional specialization from the epithelial repertoire . . . . . In the absence of stroma, epithelia show, in general, a limited capacity for survival and cytodifferentiation, failure to undergo morphogenesis, and loss of histotypic organization."

Just as the outcome of tissue development is a consequence of relations among cells and support structures (especially the ECM) and not just the differentiation of the cells themselves, it is reasonable to ask whether the maintenance of the architecture of replicating complex adult tissues is a function of relations among cells rather than a function of the individual cells themselves. This question applies to the normal maintenance of epithelia, the repair of epithelia in the face of acute (*e.g.*, wounding) and chronic (*e.g.*, ulcerative colitis) damage, and the generation of new tissues (*e.g.*, angiogenesis). This last has been the focus of considerable attention,[34–37] but adult epithelial microarchitectural integrity has been much less considered.

Fidelity and Loss of Fidelity of Microarchitecture

There is overwhelming empirical evidence from all living organisms that microarchitectural consistency of tissues is the norm. Disruption of tissue architecture—gross or subtle, acute or chronic—is part of the definition of many diseases, regardless of whether

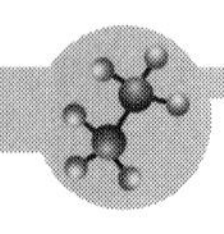

they are initiated by endogenous processes or exogenous agents. Apart from inflammatory/infectious destruction of tissue architecture, the most obvious forms of distorted microarchitecture are metaplasias and transdifferentiation.

Metaplasia involves loss of control of a whole differentiation sequence. Metaplasias used to be described regularly and in detail in histopathology reports,[38] particularly in relation to cancer, but they seem to be much less the focus of contemporary reports. Whether this is because the existence of metaplasia is taken for granted and is not seen to need explanation or because its meaning in the context of the more significant neoplasia is not understood, remains unclear. There are some obvious exceptions, perhaps most notably BE, where the metaplastic change is an established malignant precursor, but where much of the molecular biology remains to be unraveled.

Slack[39] points out that metaplasia represents a single step from normal, otherwise one would sometimes find examples of metaplasia within metaplasia. Slack further argues that the switch that is altered in any specific case must be a master switch that allows different pathways to differentiation in different tissues, for example, a homeobox gene. Again the key implication for the hypothesis being advanced here is that the cells of origin of this signal must be outside the metaplastic epithelium.

Although Barrett esophagus (BE) is the most extensively studied of the metaplasias (largely because it is a premalignant condition), there are a large number of other identified metaplasias. These include intestinalization of

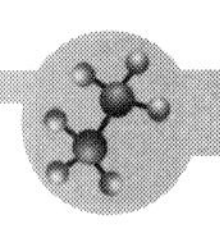

the gastric mucosa [also a cancer precursor.[40, 41]] cystitis glandularis of the bladder,[42] a variety of abnormalities of the female reproductive tract epithelium,[43] and squamous metaplasia of the lung and other sites.[44] The appearance of bony metaplasias in tissue displaying chronic inflammation is also well described.[38]

The fact that these and other metaplasias occur repeatedly in a consistent histotypic form and the overall rarity of metaplasia in general, both argue for a system whereby tissue architecture has complex controls, inasmuch as even when the controls fail, they do so in a relatively organized fashion. The tendency of the upper gastrointestinal tract, for instance, to produce lower intestine-like features may be evidence of a default state that has lost the fine-tuned structure that matches esophageal function; alternatively, it may represent a specific shift in the tissue structure that, as with normal colon, protects proliferating esophageal cells, which are not normally exposed for any duration, from environmentally induced DNA damage by burying the replicating cells at the base of crypt-like structures.[7]

Another abnormality that is even less common is the combination (sometimes called collision) tumor. These tumors exhibit mixed malignant epithelial and malignant stromal elements, *e.g.*, carcinosarcomas of the endometrium.[45] Their structure has provided a variety of explanations, but the most recent evidence suggests that they are combination tumors and that the malignant stroma and malignant epithelium arise from a single precursor.[46] These tumors demonstrate unequivocally that transdifferentiation from stromal to

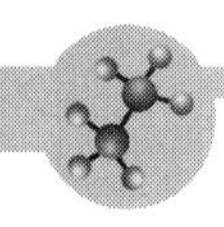

epithelial phenotype or *vice versa* can occur. Similar evidence of transdifferentiation is provided by gliosarcomas.[47] There is a great deal of other data, particularly from the developmental biology literature, that show that both epithelia and mesenchyme are capable of transdifferentiation.[48, 49] It can occur even in adult tissue in association with stromal disruption.[50]

In summary, the rarity of these phenomena (metaplasia and transdifferentiation) argue that there are controls to prevent such tissue restructuring, but the fact that the phenomena occur demonstrates that these controls can be disrupted.

Maintenance of Tissue Architecture

There is a close relationship between the maintenance of the specific architecture of tissues and the original construction of those tissues. In morphogenesis, sets of cells are defined in the embryo by a variety of organizer genes that specify the anteroposterior axis, the dorsoventral axis, and, in *Drosophila*, segments and parasegments, ultimately resulting in blocks of cells with specific relations to each other and reduced potential for differentiation.[13, 51]

Within these blocks, further separation and differentiation occurs, finally resulting in specific functional groups of cells in clearly defined structural compartments. In specific cases, the details of the control of the latter part of this process remain to be delineated, but a key element in the process is the morphogen gradient, a specific signaling molecule that determines or influences cells' fates by virtue of differences in its

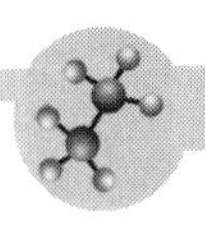

concentration, which are largely specified by the diffusion distance from its source.[52–54]

This process can occur over a continuous period of time or discontinuously with intervening growth periods (*e.g.*, in insects and amphibia). For mammals, particularly humans, there is an extensive delay between embryonic morphogenesis and pubertal morphogenesis that modifies primary and secondary sexual characteristics.

The lifelong growth, maturation, and differentiation in adult vertebrate epithelia is most like this stage of pubertal morphogenesis. The major blocks and compartments of the body design are long since established, but the final steps are a continuous task. Whether the pubertal maturation of breast, prostate, larynx, endometrium, and gonads, in part, also parallels some of the earlier morphogenetic steps, does not pose any major problems for the general argument being advanced here. It might, however, suggest that even controllers of the earlier stages of morphogenesis could be involved in the control of maturation/differentiation and thus could be potentially involved in the dysregulation (and even carcinogenesis) of these organs; that is, known morphogens may continue to exert control over the architecture of the adult tissues throughout life.

From the Concept of Morphogenesis to the Concept of Morphostasis

The major distinction between the way in which unicellular and multicellular organisms function as whole animals is the distribution of tasks among different

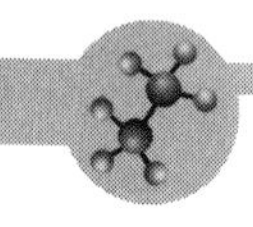

units (cells). One of the consequences of this specialization is that there are specific cell layers that constitute the interfaces between the organism and the environment. Such layers line the respiratory tract, the digestive tract, the renal tract, and the reproductive tract as well as making up the skin.

There is a special set of problems for both developing and adult epithelia that is related to the need for regular migration from basal layer to the surface, perhaps with specific functional (*e.g.*, secretory) intermediate stages or alternative fates. The control of this process will include variation in short-range signaling, cell adhesion, and cell-cell and cell-matrix interactions. At the DNA level, the cell will have several functional stages: (a) reproductive cell basally; (b) functional capacity in the intermediate stages; and (c) quiescent/senescent cell or cell remnant at the surface. Therefore, gene expression will vary considerably. This again suggests the need for a morphogen gradient or at least some method of relating position to function.

In long-lived animals, the steps of morphogenesis are insufficient because interfaces need, in addition to the initial construction phase, prolonged maintenance, repair, and renewal. As a clear distinction, there are about 1000 cells that have to last a lifetime of 10 days in *Caenorhabditis elegans*[55] compared to perhaps 10^{13} cells that persist (with replication) for 7–8 decades in a human. In lower animals, growth continues (perhaps in separate stages, as in arthropods) until death; in these animals, growth and tissue shaping are coupled. In fish and reptiles, slow overall growth persists until death, but there is little change in the relations among body parts. In birds and mammals, growth ceases at a specific time. Sexual maturation, particularly in primates and especially in humans, is dissociated in time from the initial extensive period of morphogenesis but also, as noted above, involves new morphogenetic changes. Nonetheless, in general in long-lived animals, tissue shaping and tissue maintenance are dissociated; in all of these phyla, the general problem of maintenance of the structural fidelity of replicating epithelial layers after cessation of morphogenesis needed to be solved.

The argument advanced here is that the mechanism for this process is a specialized version of tissue design machinery, the purpose of which is not to build new structures (morphogenesis) but to maintain the shape of existing structures (morphostasis). Here we advance both general and specific evidence for the existence of morphostasis and morphostats . . .

References

3. Rubin, H. On the nature of enduring modifications induced in cells and organisms. Am. J. Physiol., *258*: L19–L24, 1990.
4. Preston-Martin, S., Pike, M. C., Ross, R. K., Jones, P. A., and Henderson, B. E. Increased cell division as a cause of human cancer. Cancer Res., *50*: 7415–7421, 1990.
5. Bissell, M. J., Hall, H. G., and Parry, G. How does the extracellular matrix direct gene expression? J. Theor. Biol., *99*: 31–68, 1982.
6. Hanahan, D., and Weinberg, R. A. The hallmarks of cancer. Cell, *100*: 57–70, 2000.
7. Potter, J. D. Colorectal cancer: molecules and populations. J. Natl. Cancer Inst. (Bethesda), *91*: 916–932, 1999.
8. Rubin, H. Cancer as a dynamic developmental disorder. Cancer Res., *45*: 2935–2942, 1985.
9. Pierce, G. B. Neoplasms, differentiations and mutations. Am. J. Pathol., *77*: 103–118, 1974.

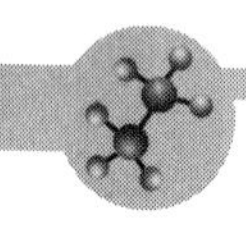

10. Pierce, G. B. Teratocarcinoma: model for a developmental concept of cancer. *In*: A. A. Moscona and A. Monroy (eds.), Current Topics in Developmental Biology, Vol. 2, pp. 223–246. New York: Academic Press, 1967.
11. Clark, W. H. The nature of cancer: morphogenesis and progressive (self)-disorganization in neoplastic development and progression. Acta Oncol., *34*: 3–21, 1995.
12. Dean, M. Cancer as a complex developmental disorder— Nineteenth Cornelius P. Rhoads Memorial Award Lecture. Cancer Res., *58*: 5633–5636, 1998.
13. Lawrence, P., and Struhl, G. Morphogens, compartments, and pattern: lessons from *Drosophila*? Cell, *85*: 951–961, 1996.
14. Perrimon, N. The genetic basis of patterned baldness in *Drosophila*. Cell, *76*: 781–784, 1994.
15. Wieschaus, E., Nusslein-Volhard, C., and Jurgens, G. Mutations affecting the pattern of the larval cuticla in *Drosophila melanogaster*. III. Zygotic loci on the X-chromosome and 4th chromosome. Roux's Arch. Dev. Biol., *193*: 296–307, 1984.
16. Nusslein-Volhard, C., and Wieschaus, E. Mutations affecting segment number and polarity in *Drosophila*. Nature (Lond.), *287*: 795–801, 1980.
17. Nusslein-Volhard, C., Wieschaus, E., and Kluding, H. Mutations affecting the pattern of the larval cuticla in *Drosophila melanogaster*. I. Zygotic loci on the second chromosome. Roux's Arch. Dev. Biol., *193*: 267–282, 1984.
18. Nusslein-Volhard, C., Frohnhofer, H. G., and Lehmann, R. Determination of anteroposterior polarity in *Drosophila*. Science (Washington DC), *238*: 1675–1681, 1987.
19. Lawrence, P. A. The Making of a Fly: The Genetics of Animal Design. Oxford, United Kingdom: Blackwell Scientific Publications, 1992.
20. Ettinger, L., and Doljanski, F. On the generation of form by the continuous interactions between cells and their extracellular matrix. Biol. Rev., *67*: 459–489, 1992.
21. Vracko, R. Basal lamina scaffold. Its role in maintenance of tissue structure and in pathogenesis of basal lamina "thickening". Front. Matrix Biol., *7*: 78–89, 1979.
22. Bard, J. Morphogenesis: The Cellular and Molecular Processes of Developmental Anatomy. Cambridge, United Kingdom: Cambridge University Press, 1990.
33. Hodges, G. M. Tumour formation: the concept of tissue (stroma-epithelium) regulatory dysfunction. *In*: J. D. Pitts and M. E. Finbow (eds.), The Fifth Symposium of the Bristish Society for Cell Biology: The Functional Integration of Cells in Animal Tissues, pp. 333–356. Cambridge, United Kingdom: Cambridge University Press, 1982.
34. Fidler, I. J., and Ellis, L. M. The implications of angiogenesis for the biology and therapy of cancer metastasis. Cell, *79*: 185–188, 1994.
35. Risau, W. Mechanisms of angiogenesis. Nature (Lond.), *386*: 671–674, 1997.
36. Hlatky, L., Tsionou, C., Hahnfeldt, P., and Coleman, C. N. Mammary fibroblasts may influence breast tumor angiogenesis via hypoxia-induced vascular

endothelial growth factor up-regulation and protein expression. Cancer Res., *54*: 6083–6086, 1994.

37. Hanahan, D. Signaling vascular morphogenesis and maintenance. Science (Washington DC), *277*: 48–50, 1997.
38. Willis, R. A. Pathology of Tumours. London: Butterworth & Co., 1967.
39. Slack, J. M. W. Epithelial metaplasia and the second anatomy. Lancet, *2*: 268–271, 1986.
40. Hamamoto, T., Yokozaki, H., Semba, S., Yasuji, W., Yunotani, S., Miyazaki, K., and Tahara, E. Altered microsatellites in incomplete-type intestinal metaplasia adjacent to primary gastic cancers. J. Clin. Pathol., *50*: 841–846, 1997.
41. Park, S. W., Oh, R. R., Park, Y. J., Lee, S. H., Shin, M. S., Kim, Y. S., Kim, S. Y., Kyung Lee, H., Kim, P. J., Oh, S. T., Yoo, N. J., and Lee, J. Y. Frequent somatic mutations of the β-catenin gene in intestinal-type gastric cancer. Cancer Res., *59*: 4257–4260, 1999.
42. Ward, A. M. Glandular neoplasia within the urinary tract. The aetiology of adenocarcinoma of the urothelium with a review of the literature. Virchows Arch. Pathol. Anat., *352*: 296–311, 1971.
43. Shevchuk, M. M., Fenoglio, C. M., and Richart, R. M. Autogenesis of brenner tumors. I. Histopathology and ultrastructure. Cancer (Phila.), *46*: 2607–2616, 1980.
44. Willis, R. A. The Borderland of Embryology and Pathology. London: Butterworths, 1962.
45. Bartsich, E. G., O'Leary, J. A., and Moore, J. G. Carcinosarcoma of the uterus. Obstet. Gynecol., *30*: 518–523, 1967.
46. Wada, H., Enomoto, T., Fujita, M., Yoshino, K., Nakashima, R., Kurachi, H., Haba, T., Wakasa, K., Shroyer, K., Tsujimoto, M., Hongyo, T., Nomura, T., and Murata, Y. Molecular evidence that most but not all carcinosarcomas of the uterus are combination tumors. Cancer Res., *57*: 5379–5385, 1997.
47. Reis, R. M., Konu-Kebleblicioglu, D., Lopes, J. M., Kleihues, P., and Ohgaka, H. Genetic profile of gliosarcomas. Am. J. Pathol., *156*: 425–432, 2000.
48. Birchmeier, C., and Birchmeier, W. Molecular aspects of mesenchymal epithelial interactions. Annu. Rev. Cell Biol., *9*: 511–540, 1993.
49. Okada, T. S. Transdifferentiation. Flexibility in Cell Differentiation. Oxford, United Kingdom: Clarendon Press, 1991.
50. Fukamachi, H. Disorganization of stroma alters epithelial differentiation of the glandular stomach in adult mice. Cell Tissue Res., *243*: 65–68, 1986.
51. Engstrom, L., Noll, E., and Perrimon, N. Paradigms to study signal transduction pathways in *Drosophila*. Curr. Top. Dev. Biol., *35*: 229–261, 1997.
52. Dyson, S., and Gurdon, J. B. The interpretation of position in a morphogen gradient as revealed by occupancy of activin receptors. Cell, *93*: 557–568, 1998.

53. Gurdon, J. B., Harger, P., Mitchell, A., and Lemaire, P. Activin signalling and response to a morphogen gradient. Nature (Lond.), *371*: 487–492, 1994.
54. Gurdon, J., Dyson, S., and Johnston, D. Cells' perception of position in a concentration gradient. Cell, *95*: 159–162, 1998.
55. Wolkow, C. A., Kimura, K. D., Lee, M.-S., and Ruvkun, G. Regulation of *C. elegans* life-span by insulin-like signaling in the nervous syatem. Science (Washington DC), *290*: 147–150, 2000.

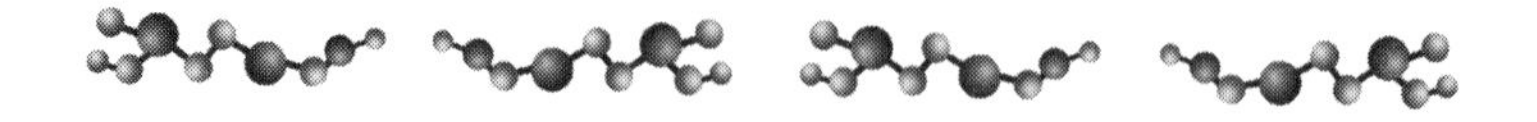

Potter goes on to cite evidence for morphostats from in vitro, in vivo, pre-cancer, and cancer studies. He suggests that epithelial cell layers may function to maintain morphogen-like gradients as developing tissue layers do. Finally, he compares the morphostat hypothesis to the mutational hypothesis of cancer formation and suggests testable predictions of the morphostat hypothesis.

Continuing along the same lines of thought, in the next article, Donald Ingber presents his ideas on cancer as a disease of tissue development. —CCF

From "Cancer as a Disease of Epithelial-Mesenchymal Interactions and Extracellular Matrix Regulation"

by Donald E. Ingber
***Differentiation*, 2002**

Cancer is commonly characterized as a disease that results from unrestricted cell proliferation. However,

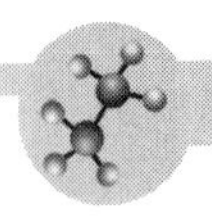

abnormal growth patterns can be observed in benign tumors, and certain normal tissues, such as bone marrow and intestine, exhibit higher cell turnover rates than seen in most cancers. The reality is that cancer is not just a disease of the cell. In addition to increased growth, classic hallmarks of malignancy include loss of normal tissue architecture, breakdown of tissue boundaries, stromal changes, angiogenesis, and compromise of distant organs through metastatic spread. Cancer therefore may be viewed to result from deregulation of the finely coordinated processes that normally govern how individual cells are integrated into tissues, tissues into organs, and organs into a functional living organism. For this reason, we must go beyond current reductionist approaches that focus on analysis of the abnormal properties of individual tumor cells. Instead, we need to examine the process of carcinogenesis in context of normal tissue formation and developmental control. Only in this way will we gain full insight into how tissues undergo malignant transformation and hence, how this process may be controlled or even reversed.

In this article, I explore cancer as a disease of tissue development. First, older (perhaps even "lost") literature will be reviewed that suggest cancer may result from deregulation of the normal process of histodifferentiation by which multiple cells collectively generate functional tissue architecture. As the large majority of cancers are epithelial in origin, this perspective leads to a view of cancer as a breakdown of epithelial–mesenchymal interactions. In the embryo, active

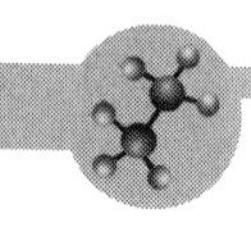

interactions between these neighboring tissues drive epitheliogenesis and determine the tissue's characteristic three-dimensional (3-D) form. Epithelial–stromal interactions also play a key role in control of angiogenesis in the surrounding stroma and thereby guide the formation of a functional vasculature that is required to feed the growing organ. As will be described, the extracellular matrix (ECM) that accumulates along the epithelial–mesenchymal interface as a result of these interactions is a critical control element in these developmental processes. More recent experimental results will be summarized that provide insight into how ECM acts locally to regulate individual cell responses to soluble growth factors and thereby control tissue patterning. A model of cancer formation will then be presented along with supporting experimental data that suggest deregulation of tissue growth and form during cancer formation may result from progressive breakdown of ECM-dependent developmental constraints.

Cancer as a Disease of Epithelial–Mesenchymal Interactions

Genesis of characteristic epithelial tissue form (e.g. acinar, tubular, branched, planar) is determined through complex interactions between the epithelium and its underlying mesenchyme in the embryo. Studies in which embryonic epithelia and mesenchyme were isolated independently of different tissues and then recombined heterotypically revealed that the precise 3-D form that the tissue will express ("histodifferentiation") is determined by the source

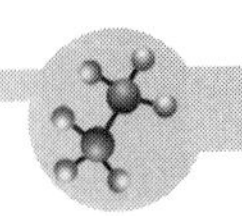

of the mesenchyme. In contrast, the epithelium governs what specialized products the cells will produce ("cytodifferentiation").

One of the key products of epithelial–mesenchymal interactions is the accumulation of a specialized ECM scaffold, called the "basement membrane" (BM) along the epithelial–mesenchymal interface. The BM functions as a extracellular complex of informative molecules that guides the differentiation, polarization, and growth of adjacent adherent cells, in addition to stabilizing the tissue's characteristic 3-D form. Localized differentials in BM turnover also play a central role in tissue patterning. The epithelium physically stabilizes tissue morphology by producing BM; the mesenchyme actively induces histogenic changes in form by degrading BM at selective sites. The highest cell growth rates are observed in regions that exhibit the most rapid BM turnover, such as the tips of growing epithelial lobules during salivary morphogenesis and in regions of capillary sprout formation during angiogenesis. At the same time, the mesenchyme slows matrix turnover and induces BM accumulation by depositing fibrillar collagen in slower growing regions of the same tissue (e.g. in clefts between growing lobules). Branching morphogenesis in mammary gland also can be either stimulated or inhibited by increasing or decreasing ECM turnover, respectively, using modulators of stromal-derived matrix metalloproteinases. Thus, the stability of epithelial tissue form depends on the presence of an intact BM, whereas changes in tissue pattern are driven by spatial differentials in BM

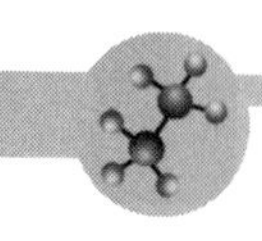

turnover. Most importantly, both activities are governed through active interactions between closely apposed epithelial and mesenchymal tissues.

Similar developmental controls are apparently sustained throughout adult life. Mature epithelial tissues retain the ability to undergo normal morphogenesis when mixed with embryonic mesenchyme and to switch histologic form when combined with different types of stromal tissue (e.g. epidermal grafts in different dermal sites). Studies of teratocarcinoma cells that exhibit multiple differentiated lineages suggest that tumors may mimic their tissue of origin in both appearance and mode of development. Given that over 90% of tumors are carcinomas (epithelial in origin), then formation of most cancers may involve deregulated interactions between epithelial cells and underlying mesenchymally derived stromal tissue. In fact, examination of the microscopic anatomy of various cancers supports this hypothesis. Examples include the finding that tumor architecture varies depending on the source of connective tissue whereas production of stromal collagen by host fibroblasts depends on the epithelial tumor cell type. The finding that an epithelial neoplasm must gain the ability to induce angiogenesis within its surrounding stroma to switch from hyperplasia to cancer and to progressively expand in size is another clear example of how epithelial–stromal interactions are critical to cancer progression.

Epithelial–mesenchymal interactions also may contribute to tumor initiation. For instance, chemical carcinogenesis of epidermis requires the presence of

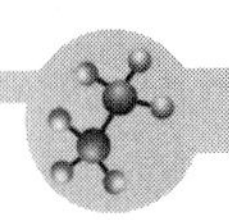

closely apposed carcinogen-treated dermis. Malignant transformation of embryonic mouse submandibular gland by polyoma virus also only occurs in intact or reconstituted glands; transformation cannot take place in isolated submandibular epithelium or mesenchyme, even though the resultant tumor is epithelial in origin. Moreover, once transformed, these epithelial tumor cells can then substitute for mesenchyme in either normal epithelial morphogenesis or viral transformation of isolated embryonic epithelia. Grafted human epithelial tumors also recruit normal murine stromal cells to become tumorigenic in nude mice. But most impressive is the finding that combination of various disorganized epithelial cancers with normal embryonic mesenchyme results in reversal of the malignant phenotype, as evidenced by restoration of normal epithelial organization and histodifferentiation. Thus, continued epithelial–mesenchymal interactions appear to play an important role in the maintenance of normal tissue architecture in adults and deregulation of these interactions may contribute to both early and late stages of cancer formation . . .

Ingber then discusses how the BM serves as a scaffold for normal tissue development, how disruption of the BM is a sign of malignancy, and how experiments showed that pancreatic cancer cells could establish normal tissue architecture when grown on BM. He then discusses how mechanical factors play a role in normal cell and tissue growth. —CCF

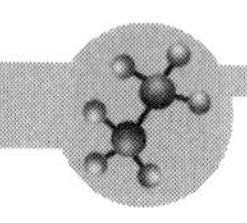

Normal Histogenesis

Twenty years ago, we proposed that ECM may contribute to control of both morphogenesis and tumor formation based on its *mechanical* properties; that is, its ability to physically resist cell tractional forces. The concept was that the local variations in ECM remodeling that are observed during morphogenesis would change ECM structure and mechanics. In this model, the local regions of BM at tips of growing epithelial glands and new capillary sprouts that become thinner due to high ECM turnover would be expected to become more compliant. All soft tissues experience isometric tension or a tensile "pre-stress" based on the generation of tractional forces by their constituent cells. Because it is under tension, a weak spot in the BM would stretch out more than the neighboring tissue, just like a "run" in a woman's stocking. This alteration in ECM mechanics would change the balance of forces that are transferred across cell surface integrin receptors that physically connect the ECM to the internal cytoskeleton (CSK) within adjacent adherent cells. Increased tension on these adhesion receptors would, in turn, promote cell and CSK distortion in these distended regions and thereby alter cell sensitivity to soluble cytokines, resulting in the localized growth that drives tissue patterning. If this increase in cell division were paralleled by a commensurate local increase in BM expansion (due to net BM accumulation), then coordinated budding and branching would result . . .

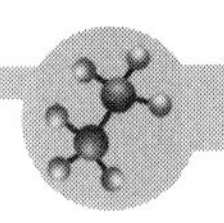

Ingber cites experimental studies that show how the substrates upon which cells are grown can influence their shape and ability to regulate growth. —CCF

Carcinogenesis

If cancer results from a breakdown of the rules that guide normal histogenesis, then loss of this form of tension-driven structural remodeling could contribute to neoplastic disorganization of tissue architecture. As described above, the BM commonly remains physically intact during early stages of tumor formation (i.e. prior to malignant transformation and invasion), but a reduction in BM thickness or subtle decreases in the levels of certain ECM constituents often can be detected. In contrast to changes observed during normal tissue development, these neoplastic changes are not restricted in space or time and hence tissue disorganization results. In fact, it is this loss of tissue pattern that usually catches the eye of the pathologist who recognizes it as abnormal. As in the embryo, continued changes in BM structure that lead to increased mechanical compliance (e.g. thinning) may promote cell distortion or increase CSK tension locally and thereby increase the sensitivity of adjacent cells to mitogenic stimuli. But in this case, a piling up of cells would result because the BM does not expand in parallel to match increases in cell number, as occurs in normal development.

If these changes are maintained over many years and the growth stimulus is sustained, cells that grow free of anchorage *in vivo* may spontaneously emerge just as continued culturing of normal cells may lead to spontaneous transformation *in vitro*. This transformation process would require that the cells gain the ability to grow independent of both ECM adhesion and cell distortion to fully overcome normal crowd controls. Natural selection and expansion of this autonomous cell would result in the "clonal" origin of proliferating tumor cells, yet the evolutionary process that led to creation of this cancer cell would have taken place at the tissue level. Cell growth and survival free of contact with the BM is then sufficient to explain the disorganization of normal cell-cell relations that is observed during subsequent stages of neoplastic transformation . . .

Ingber then cites experiments in which normal and transformed cells are grown on various substrates that restrain cell spreading and, therefore, maintain cell differentiation. He postulates how mechanical factors can be linked to changes in the cell's biochemistry. —CCF

Mechanical Control of Normal and Malignant Tissue Differentiation: An Overview

The most important point of this discussion is that cancer represents more than uncontrolled cell growth; it is a disease of tissue structure that results from a breakdown of normal epithelial–mesenchymal interactions. The structural coordination, homogeneity of cell form, and intercellular communication required for successful tissue

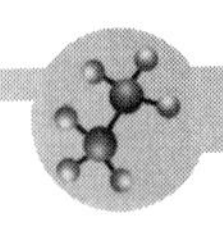

function are maintained by normally constant architectural relationships. In the tension-driven remodeling hypothesis for developmental control presented here, local thinning of the BM scaffold (which resists cell tractional forces and stabilizes tissue form) locally increases CSK tension within the adjacent epithelial cells. Due to the use of tensegrity for control of shape stability, this local change in the mechanical forces balanced across integrins would produce cell and CSK distortion that, in turn, would alter cellular biochemistry and increase cell growth. As long as accelerated cell division was matched by a commensurate increase in BM expansion (net BM accumulation), then orderly tissue expansion and morphogenetic remodeling would proceed.

In certain situations in which epithelial–mesenchymal interactions become deregulated, accelerated BM remodeling may lead to a continued release of mechanical constraints and an associated increase in cell growth without commensurate BM extension (i.e. no net BM accumulation). Cancer formation would be prevented as long as cell viability and proliferative capacity remained dependent upon continued anchorage to ECM. Thus, this hyperplastic state would be reversible; if the stimulus ceased, cells no longer in contact with the BM would undergo apoptosis and normal tissue form would return. On the other hand, if the conditions that led to release of tensile constraints within the tissue were sustained over an extended period of time (years), then this continued stimulus for cell division may lead to selection of an anchorage-independent population that, by definition, could proliferate

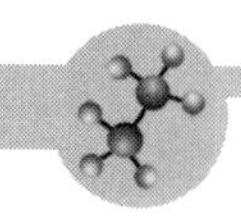

autonomously. As the tumor grew in size, autonomous epithelial cells would become separated from the stroma by large distances and so would become less susceptible to the normal regulatory influences of the mesenchymally derived connective tissue. Loss of normal cues from the deregulated epithelium also may alter stromal cell behavior and further compromise ECM regulation.

In this manner, a positive feedback system may develop that would move the tissue along a spectrum of progressive deregulation and eventually result in invasion of epithelial cells through the BM (i.e. malignant conversion). This may be manifested either through increased BM degradation relative to synthesis, or through acquisition of some new transformed cell product that in some way further compromises ECM-dependent developmental control. While progressive BM dissolution and altered CSK structure may be directly involved in early carcinogenesis in certain tumors, other cancers may enter this positive feedback loop at a later stage after gaining the ability to proliferate independently of anchorage by chemical, genetic or viral means. In this manner, gene mutations for growth signaling, adhesive (integrin, cadherin) signaling, and mechanical (cell shape) signaling may all need to occur for full malignant conversion of a benign neoplasm.

It is important to note that the changes in the epithelial BM that are observed during neoplastic transformation of the epithelium are quite similar to those induced within the BM of nearby capillaries when they are induced to grow by epithelial tumor-derived angiogenic factors. Tumor angiogenesis is required for progressive growth and expansion of the tumor mass. Thus, altered epithelial–stromal interactions that compromise ECM regulation in the local tissue microenvironment may actively contribute to all stages of cancer development, including early tumor initiation, the onset of malignant invasion, and the final switch to the angiogenic phenotype that represents the end of tumor dormancy . . .

Finally, Ingber suggests how the work on mechanical control of cell division and tissue development by the ECM can lead to the development of in vitro models of cancer, to therapies targeting biochemical signals (integrins) related to the mechanical control, and to a whole systems approach to studying cancer. —CCF

4 Artificial Modes of Differentiation

Each year, many people suffer from trauma or disease that damages internal organs. Heart disease weakens the muscle, causing it to fail. During an equestrian accident, actor Christopher Reeve of Superman fame suffered a spinal cord injury that rendered him paralyzed and incapable of breathing on his own. Hepatitis damages the liver beyond its regenerative capacity. Diabetes causes kidney disease and renal failure. The most prevalent treatment for these diseases is organ transplant; however, the supply of donor organs does not meet the demand, and the waiting lists are long. Often, donor organs are rejected by the recipient's immune system or the recipient must take immune-suppressing drugs for a lifetime, which renders them susceptible to common infections.

What if we could manufacture organs at will? In an episode of Star Trek: The Next Generation *entitled "Ethics," Lieutenant Worf*

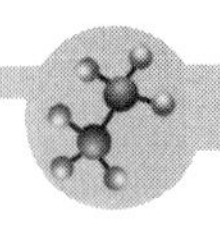

suffers a spinal cord injury that paralyzes him. A visiting scientist tells of her research on making new tissue to treat such injuries. Reluctantly, Worf agrees to the procedure, a new spinal cord is manufactured, and Worf is ultimately cured. Is this just science fiction? Scientists are currently working on ways to grow new skin to treat burn victims. This is but one example of tissue engineering. In the following article, David Williams describes the current status of the field. —CCF

From "Benefit and Risk in Tissue Engineering"

by David Williams
***Materials Today*, May 2004**

Tissue engineering is a radically new concept for the treatment of disease and injury. It involves the use of the technologies of molecular and cell biology, combined with those of advanced materials science and processing, in order to produce tissue regeneration in situations where evolution has determined that adult humans no longer have innate powers of regeneration. Tissue engineering, however, along with some other aspects of regenerative medicine such as gene therapy, has yet to deliver real successes in spite of a considerable science base and investment in the commercial infrastructure. This article addresses the underlying issues of benefit and risk in tissue engineering in an attempt to understand why this situation has developed.

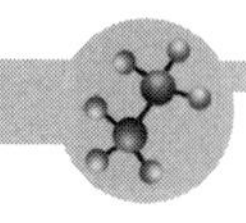

There is an inherent, virtuous logic to tissue engineering that sounds too good to be true. By my definition, tissue engineering is the persuasion of the body to heal itself, achieved by the delivery to the appropriate site of cells, biomolecules, and supporting structures.[1] It specifically involves the regeneration of new tissue to replace that which has become diseased or injured, the significance of which is that we, as adult humans, do not normally possess this ability. We may repair ourselves under some very limited circumstances (for example, bone fractures and injured skin may undergo repair) but, even when this does occur, this usually involves nonspecific reparative tissue (i.e. scar tissue) rather than the regeneration of the specific functional tissue that has been affected.

The essence of tissue engineering is that those cells capable of initiating and sustaining the regeneration process are "switched on," perhaps through growth factors or genes, so that they generate new functional tissue of the required variety. This may be achieved with the help of a scaffold or matrix to guide the geometrical or architectural shape of the new tissue and may take place on a customized basis at the site of the injury in an individual patient or on a more industrial scale in an *ex vivo* bioreactor, where the resulting tissue construct is re-implanted into the patient.

This all sounds relatively straightforward but, of course, it is not. In spite of a massive investment in the underlying science and scale-up procedures, very few clinical conditions are being currently addressed by tissue engineering, and commercial success has been very hard to achieve. This article addresses the underlying

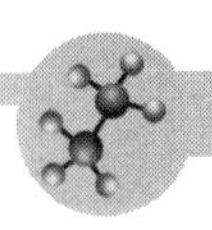

factors that are encountered in the harsh reality of tissue engineering. In doing so, the actual and projected applications are mentioned first, followed by an analysis of the gaps in our scientific knowledge and a discussion of the clinical, regulatory, ethical, and commercial risks.

Potential Clinical Opportunities

Whenever a new concept of medical therapy becomes available, it is extremely important for it to be applied to the most relevant diseases and conditions, where the expected benefits match as closely as possible the requirements of patients. For a radically new concept such as tissue engineering, this represents quite a challenge. On the one hand, there are very good prospects that tissue engineering could address medical conditions for which there are no existing successful therapies and it could be argued that it is on these conditions where the concept should be focused. On the other hand, these applications could be associated with high risks and the emerging area, including the commercial infrastructure, could be severely damaged if attempts at such high-profile, high-risk conditions end in highly publicized failure. The whole area of gene therapy, a similar emerging technology within regenerative medicine, suffered such a setback a few years ago with deaths in the USA and France.[2] In describing potential clinical opportunities, therefore, it is necessary to consider situations of wide-ranging clinical benefits and risks.

The two types of tissue most commonly considered in tissue engineering products and processes are skin and cartilage. As noted earlier, skin is able to be repaired and to regenerate itself to some extent. It is obvious, however, just by looking at an area of repaired skin that the resulting structure, although providing a functional barrier between the internal tissues of the body and the outside world, does not look nor feel like natural undamaged skin; indeed, it is usually scar tissue of quite different texture and appearance. Moreover, there are many conditions where it is impossible for skin to repair or regenerate sufficiently well to give a clinically acceptable outcome. The two main conditions here are where massive amounts of skin have been damaged, as in burns injuries, and in those situations where there are irreversible changes to the underlying vascularized tissue such that the skin is deprived of its nutrition source, creating an ulcer. A decubitus ulcer in the elderly (e.g. bed sores) or the common foot ulcer of the diabetic patient are good examples. The first generation of commercially available tissue engineering products have attempted to address these areas of skin regeneration, with some degree of success.[3]

With cartilage, a relatively simple three-dimensional connective tissue that has no intrinsic powers of regeneration, there are a number of conditions where such regeneration could be of enormous benefit clinically. The articular cartilage of joints such as the hip and knee suffers both degenerative diseases, such as osteoarthritis, and trauma, for example in sports injuries. A great deal of attention has been paid to the replacement of diseased joints over the last several decades and medical device technology has produced a

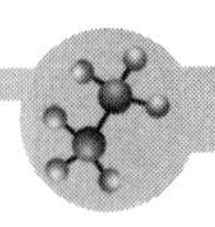

very effective portfolio of procedures and devices that are expected to give successful replacement of these joints in 90% of patients for at least ten years.[4] It may well be that tissue engineering will enable the regeneration of diseased joints in the future, but this is not seen as either technically feasible nor economically viable at this stage. Of more importance in tissue engineering is the possibility of faster, effective treatment of small lesions arising from sports injuries, where both of these technical difficulties are more easily overcome and the economic equation is reversed. In addition, one of the degenerative conditions in the body that is becoming increasingly common and significant, from both individual quality of life and socio-economic points of view, is that of the spine, especially intervertebral disks. These are complex heterogeneous anisotropic structures composed mainly of cartilage and fluid, damage to which is debilitating and difficult to repair effectively and permanently. Tissue engineering is widely seen as a potential vehicle for disk regeneration.

The spinal column is, of course, crucially important from the perspective of the spinal cord and it is well known that serious damage to this cord of nervous tissue results in permanent disability. Nerves, whether of the central nervous system (i.e. brain and spinal cord) or the peripheral nerves, are notoriously difficult to repair. When soft tissue is cut, the very small nerves within it may regenerate such that, for example, sensation gradually returns to tissue after an operation, but there is a serious limitation to this and many injuries to nerves in the arms and legs are impossible to treat effectively.

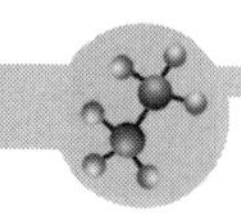

Nerve cells can grow under some conditions to produce a degree of function, but they usually need considerable help and guidance to do so. Tissue engineering is addressing these problems by the use of nerve guides and growth factors in order to stimulate the timely and effective regeneration of nerves.[5] Some see this as potentially the most important area of tissue engineering, involving as it does the introduction of a new therapeutic modality where none other exists. Perhaps of even more significance than injury to the central nervous system (CNS) are the many types of neurodegenerative disease, of which Alzheimer's and Parkinson's are among the more common. These are often discussed as target areas for tissue engineering and other regenerative medicine procedures, although precisely how this could be achieved is a matter of debate.

There is insufficient space here to discuss all the potential applications of tissue engineering as it could be argued that most noninfectious and noncancerous diseases, especially those involving congenitally absent or damaged tissues, physically traumatized tissue, and degenerative or metabolic conditions in the aged, are potential targets. It is appropriate to use as a final example those conditions that affect the central circulatory system, since these are the major cause of morbidity and mortality in first world countries. These include heart failure, heart valve disease, and atherosclerosis in the major circulation. They epitomize the dilemma of the scientific and policy makers in tissue engineering. Taking heart valve disease as an example, this represents a group of conditions that are life-threatening but declining rather than increasing in incidence and for which there are effective, although not perfect, treatment modalities, involving implantable medical devices, already available and well proven. As noted in a later section, there will be profound risks involved with the use of a customized tissue-engineered heart valve in patients such that, taking into account the availability of alternatives, the relative benefits may never outweigh the risks, except possibly in some pediatric cases where current heart valve prostheses are not always effective in the long-term. On the other hand, with heart failure itself, which may involve extensive damage to the heart muscle through a lack of blood supply following blockage of one or more coronary arteries, effective treatments are not really available and the potential benefits of regeneration of the myocardium (the muscular tissue of the heart) are immense. It has been demonstrated that the damage to the myocardium associated with an infarction (a "heart attack") may be reversible[6] and the delivery of the all-important cardiomyocytes (the cells of this tissue) through a tissue engineering approach has considerable potential.[7]

The potential clinical benefits are wide ranging and very attractive. Before discussing the aspects of risk that have to be set against these benefits, it is instructive to consider briefly the state of the science that underpins tissue engineering in order to see how realistic all of this is.

The Status of Tissue Engineering

It may be obvious from the above discussion that there are many potential methodologies by which the tissue

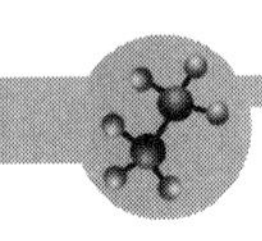

engineering concept can be translated into clinical practice. It is possible, however, to identify a central tissue engineering paradigm that describes the essential route from individual cells to regenerated tissue. This route involves the phases of cell sourcing, cell manipulation, cell signaling, tissue expression, possibly within a bioreactor, the implantation of the tissue construct into the host, and its full, effective incorporation into that host.

As far as the source of the cells is concerned, there are many challenges, including the most important question of all in tissue engineering. Leaving aside, for the moment, the use of animals as sources of cells since this practice, as part of xenotransplantation, is effectively banned throughout the world because of ethical and disease transmission issues, we have two potential sources of cells: those from the patient and those from a donor. In the former case, autologous cells, as they are described, could be the fully differentiated cells of the tissue in question, that is chondrocytes to produce cartilage, osteoblasts to produce bone, glial cells to produce nerve tissue, myocytes to produce muscle, and so on. Alternatively, they could be the patients' stem cells, derived from their own bone marrow or possibly their circulating blood, which have been persuaded to differentiate into the required cells, i.e. changed from stem cells into the required chondrocytes, osteoblasts, and so on. An interesting alternative for the future here involves the harvesting of cord blood when a baby is born, which is then stored at low temperature until required for a tissue engineering procedure later in life. Cells from donors are referred to as allogeneic cells,

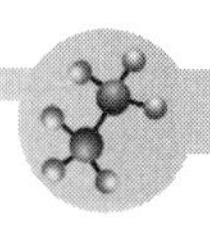

which could be from a known individual, but are more likely to be obtained from a commercial cell bank, where cells from an original donor are cultured, sorted, and expanded to produce a readily available supply of standard quality cells of the required type. These cells, and the tissue they produce, are, of course, foreign to the eventual hosts of the tissue engineering products to which they give rise. An immensely important factor here is the possibility of using allogeneic stem cells, which could, in theory, be adult, fetal, or embryonic. There are many who would argue that embryonic stem cells offer the greatest potential here, but there is a current moratorium on the use of embryonic stem cells for use in patients. While there are strong demands for this to be made permanent with respect to reproductive technologies, there are good arguments in favor of strictly controlled use of embryonic stem cells for regenerative therapies, as noted very recently with the successful use of such cells in the cloning of human cells with potential in these therapies.[8]

The issues around cell manipulation and signaling can be treated together. These concern the conditions in which the cells are grown, such that they receive the right nutrients and signals to adopt and/or retain the required characteristics (referred to as the phenotype of the cell) for the desired endpoint. This is at the heart of the scientific basis of tissue engineering, since these cells have to be stimulated to produce the required tissue because they are not intrinsically able to do so. The signals are of two varieties, molecular and mechanical. Molecular signals are usually, but not necessarily, growth

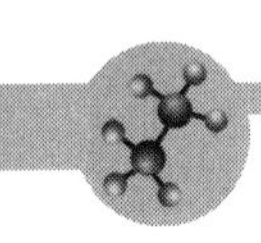

factors. A combination of the most appropriate factors and their concentration profile over time provide the essential key to cell behavior. It is important to note here that many cells derived from normal adult human tissue may not be able to respond to such growth factors sufficiently well in normal culture to be of any practical value. This may be addressed in two ways, with an additional molecular signal or by mechanical signals. The molecular signal may take the form of a gene, which can be inserted into the cells to change some aspect of their character and increase the efficiency with which they carry out certain functions. The performance of chondrocytes in culture can be radically altered by the transfer of so-called SOX9 genes, for example.[9]

It is with mechanical signaling that bioreactors and material scaffolds are introduced. Cells do not normally function in isolation or, indeed, within diffuse collections of cells alone. They are found within an extracellular matrix and their behavior is regulated by signals that are passed between cells and their matrix. One of the more important elements of this signaling between cells and the matrix involves mechanical forces and the behavior of cells is often dominated by the nature of these forces through a phenomenon known as mechanotransduction.[10] In tissue engineering processes, forces can be applied to cells through a fluid medium or by structural forces delivered by a substrate. The fluid medium may be contained within a culture vessel, which is known as a bioreactor.[11] In the vast majority of tissue engineering processes, cells are seeded into a porous scaffold within which they are provided with the molecular

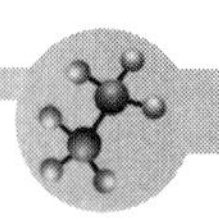

signals discussed above. In addition, their interaction with the scaffold material allows for the transmission of the forces from those surfaces. The vast majority of scaffolds are made from biodegradable materials, the degradation mechanisms of which are crucial in determining the outcome of tissue regeneration.

Once the required volume of tissue has been generated through the activity of these cells, it may be implanted into the recipient. This is not a trivial point, since there has to be complete and effective incorporation, which implies integration of the new tissue within the vascular and nervous system of the host, especially the development of the optimal blood supply through a process known as angiogenesis[12] and the control of any inflammatory or immune system responses.[13]

Bearing in mind the complexity of all of these issues, it is not surprising that, although considerable progress has been made in a short period of time, there are still many scientific issues to resolve. Among the most important of these are the questions that surround autologous cell expansion, the maintenance of their phenotype, and the optimization of their efficiency through gene transfection, the development of effective nonviral vectors for gene transfection, the control of differentiation of stem cells in the abnormal environment of bioreactors, the control of tissue regeneration in cocultured heterogeneous anisotropic systems, the optimization of mechanotransduction, the procedures of immunomodulation with allogeneic cell-derived products, the optimization of vascularization and angiogenesis, the control of inflammation

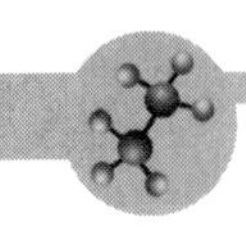

during incorporation into the host, and the determination of the functionality of regenerated tissue . . .

REFERENCES

1. Williams, D. F., *The Williams Dictionary of Biomaterials*, Liverpool University Press, Liverpool, (1999)
2. Dobbelstein, M., *Virus Res.* (2003) 92 (2), 219
3. Falabella, A. F., et al., *Arch. Dermatol.* (2000) 136 (10), 1225
4. *Guidance on the selection of prostheses for primary hip replacement*, 2, National Institute for Clinical Excellence (www.nice.org.uk), UK, (2000)
5. Schlosshauer, B., *et al.*, *Brain Res.* (2003) 963 (1-2), 321
6. Cleland, J. G. F., *et al.*, *Eur. J. Heart Failure* (2003) 5 (3), 295
7. Shimizu, T., *et al.*, *Biomaterials* (2003) 24 (13), 2309
8. Cloned human embryos are stem cell breakthrough, *New Scientist*, Feb 12, 2004
9. Kolettas, E., *et al.*, *Rheumatology* (2001) 40 (10), 1146
10. Ingber, D. E., *Ann. Med.* (2003) 35 (8), 564
11. Engelmayr, Jr., G. C., *et al.*, *Biomaterials* (2003) 24 (14), 2523
12. Zisch, A. H., et al., *Cardiovasc. Pathol.* (2003) 12 (6), 295
13. Kirkpatrick, C. J., *et al.*, *Biomol. Eng.* (2002) 19 (2-6), 211

A key aspect of tissue repair or engineering is the use of undifferentiated cells, usually of embryonic origins, that are subsequently enticed chemically or genetically to differentiate into specialized tissue. For example, many people suffer from Parkinson's disease, a debilitating disease in which certain neurons in the brain stop producing an important neurotransmitter called dopamine. The patient suffers numerous tremors, loss of motor control, and eventually death. If neurosurgeons transplant fetal brain

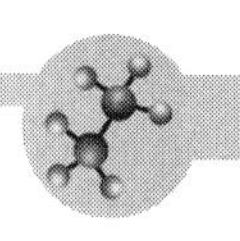

cells into the appropriate area, the stem cells will produce dopamine and minimize or eliminate the Parkinsonian symptoms. In another example, type I diabetic patients lack insulin-producing cells in the pancreas; the lack of insulin disrupts glucose homeostasis, thereby leading to many problems. Stem cells can be transplanted into the pancreas and induced to form cells, which will secrete insulin and correct the diabetes.

Despite potential medical successes, the use of embryonic stem cells remains controversial. Many embryos that provide the stem cells were generated by assisted reproductive techniques and would otherwise be discarded. However, as these embryos represent potential humans, do investigators have the moral and ethical rights to use them? Dr. Christopher Cogle and his colleagues address these issues in the following article. —CCF

From "An Overview of Stem Cell Research and Regulatory Issues"

by Christopher R. Cogle, M.D., Steven M. Guthrie, B.S., Ronald C. Sanders, M.D., William L. Allen, J.D., Edward W. Scott, Ph.D., and Bryon E. Petersen, Ph.D.
***Mayo Clinic Proceedings*, 2003**

Embryonic Stem Cells

Some stem cells have a greater capacity of self-renewal and multilineage differentiation than others. At the time of conception, the fertilized egg (zygote)

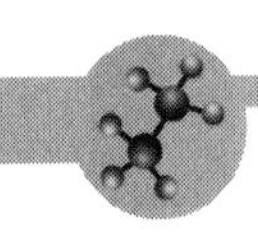

contains dividing cells (blastomeres) that form an embryo and placenta. These blastomeres are *totipotent*; they have the potential to form an entire living organism. After about 4 days, these totipotent cells begin to specialize and form into a hollow ball, the blastocyst, containing an outer shell (trophoectoderm) and a cluster of cells called the inner cell mass (ICM), from which the embryo develops. Human embryonic stem (hES) cells are derived by removing the trophoectoderm, which would normally become the placenta, and culturing cells from the ICM.[6] These hES cells of the ICM are *pluripotent*; they are able to differentiate into tissues of all 3 germ layers but cannot produce another embryo because they are unable to give rise to the placenta and supporting tissues. Transplanting hES cells into a woman's uterus would not produce a fetus. Blastocysts harvested for hES derivation are usually acquired from unused in vitro fertilizations (IVFs). The hES cells express specific surface antigens, as well as OCT-4 and human telomerase, proteins associated with a pluripotent and immortal phenotype.[7] A second source of pluripotent cells is human embryonic germ cells, derived from the fetal gonadal ridge, which normally gives rise to either sperm or egg.[8] It remains to be tested whether human embryonic germ cells have the same capacities that make hES cells so important.[9] A hallmark of hES cells is their long-term self-renewal capability in culture dishes. Whereas adult stem cells divide asynchronously and eventually lose their ability to self-renew, hES cells have been cultured over many

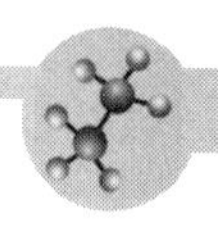

years, maintaining developmental potential, proliferative capacity, and karyotypic stability.[10]

Because of this unlimited self-renewal and boundless developmental potential, hES cells may serve as powerful tools to unlock important medical challenges. As a basic science tool, these cells help to identify molecular mechanisms in pluripotent cell differentiation, affording investigators a much better understanding of fetal development.[11] Ultimately, this understanding aims to reduce infertility, pregnancy loss, and birth defects—major health care challenges in the United States. Also, hES cells are helpful in drug development, identifying potentially embryotoxic teratogens.[12] Moreover, normal cell lines derived from these pluripotent cells may serve as representative tissues for in vitro toxicity testing of medicinal compounds in development. Finally, organ regeneration from hES cells would not only halt disease progression but also could help to remodel damaged organs. Investigators are already using embryonic stem cells to create heart muscle, brain, pancreatic islet cells, and blood vessels.[13–16] Techniques used for tissue engineering include transplanting a patient's somatic cell nucleus into an enucleated oocyte, activating the cell to mimic fertilization, culturing the totipotent cells in a dish, and then differentiating the cells into tissue of need. The resulting tissue, whether cardiac, pancreatic, hematopoietic, neural, or hepatic, is genetically identical to the patient's tissue, would not be rejected because of identical HLA antigen expression, and could be used for tissue repair . . .

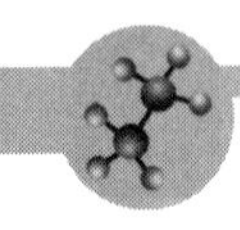

Adult Stem Cells

It is generally accepted that each organ of our body is in balance between degradation and repair. The liver that we were born with is not the same liver that we have when we die. Throughout life, toxic insults wound our organs, bringing about the question of what keeps the balance between destruction and construction. In adults, stem cells have been found in many tissues, such as liver, bone marrow, pancreas, and brain, maintaining this homeostasis. Moreover, some of these adult stem cells, once thought to mend only local property, also help in disaster relief of other more distant organs.[37–38]

To test cell plasticity, investigators use a variety of cell transplantation models. The basic procedure includes injecting donor cells of interest into a recipient and subsequently analyzing the recipient's organs for donor contribution. Modifications have been made to this basic procedure, sometimes resulting in conflicting reports in stem cell plasticity.[56, 57] These differences in findings may be explained by disparities in donor stem cell separation, cell cycle of transplanted cells, time from transplantation to evaluation of end-organ chimerism, type of injury eliciting plasticity, and ability of target niche to support stem cell transdifferentiation.[58]

Thus, before plasticity can be confirmed, several criteria must be addressed. First, clonal repopulation should be demonstrated from the transplanted stem cell. To assert plasticity potential from a specific cell type, transplantation should not be performed with a mixture of undefined cells. In experimental animal systems, researchers are now using single cell transplant and retroviral stem cell tagging to test clonal plasticity.[38, 56, 59, 60] Second, a self-renewing cell should be responsible for observed plasticity. To address this challenge, hematopoietic stem cell (HSC) investigators transplant bone marrow from the first transplant recipient into a secondary transplant recipient. If donor reconstitution develops in the secondary transplant recipient, a self-renewing stem cell is present and active. Although cells capable of organ regeneration have been identified in various organs, if the cell does not exhibit self-renewal, it is not considered a stem cell; rather, it is termed a *progenitor* cell. Third, multiple markers and proper morphology should be used in tissue analyses. Recently, stem cells thought to transdifferentiate into pancreatic islet cells were found to have simply absorbed insulin from the tissue culture media.[61] Relying on single markers to follow cell fate can be misleading. Fourth, the transdifferentiated cells should be functional. Grant et al[38] recently showed that the adult HSC can make blood and blood vessels. To show functional plasticity, they found that the donor-derived blood vessels contained blood coursing through the vessel lumens. Fifth, robust repopulation is preferred. Some investigators have scrutinized adult stem cell plasticity based on observations that small collections of cells cannot repopulate whole organs.[56, 57] Although robust repopulation is an ultimate end point in working with adult stem cells, the first step is to test whether the phenomenon exists and then identify

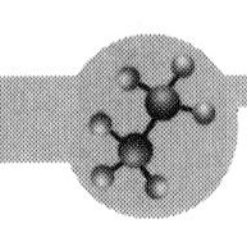

mechanisms to increase donor engraftment. Sixth, analyses for cell fusion between stem cell and differentiated cell should be performed. In culture dishes, stem cells can acquire the markers of differentiated cells by fusing with them spontaneously.[62, 63] Some investigators regard cell fusion as an artifact confounding plasticity; however, many organs in the body, including the liver, heart, skeletal muscle, and brain, have functional multinucleated cells. Until proved otherwise, fusion may be a natural and useful process . . .

Concerns Regarding the Use of Stem Cells

Certainly, stem cells are not the first human discovery to stretch the boundaries of medical knowledge and create waves of ethical debate. Since ancient times, society has admonished man for approaching these boundaries, eg, the Greek myth of Icarus who did not heed his father's command; he reveled in the "unnatural" sensation of flight and then plummeted to his death after the sun melted his wings. This Greek myth embodies our apprehensions about meddling with nature. Galileo Galilei, who stretched the boundaries of astronomy and posited that the Earth rotates on its axis and revolves around the sun, was eventually condemned for heresy. In more contemporary times, our society has grappled with permission to perform autopsies for crucial understanding of human anatomy and consent to produce recombinant DNA for lifesaving medications. In all instances of stretching knowledge boundaries, a societal consciousness was at play, in some ways

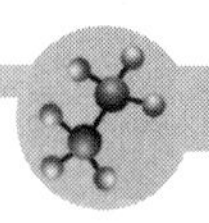

encumbering progress and questioning techniques of intervention.

Additionally, shared concerns among all instances of testing medical boundaries, as well as using stem cell technology, include issues of safety, efficacy, and resource allocation. For decades, patients have undergone adult HSC transplantation in the treatment of immune deficiencies and cancer. Although graft-vs-host disease and post-transplantation infections are major risks of allogeneic transplantation, investigators have worked to minimize these consequences, and many patients accept these risks in the hope of the lifesaving benefit of disease eradication. In contrast, reproduction by SCNT into embryonic stem cells has been inefficient and carries the concerns of developmental abnormalities and early aging.[22, 23, 32–36] However, the field of stem cell therapy is still in its infancy, with researchers incrementally improving safety, efficacy, and applicability to a wider spectrum of disease.

Stem cell therapy differs from previous technologies in how these founts of plasticity are tapped. Adult stem cells are typically acquired by harvesting adult tissues. Patients give informed consent and usually undergo little risk at donation. In contrast, hES cells are obtained by culturing cells from the ICM of a blastocyst, usually acquired from an unused human embryo produced by IVF or from an already aborted fetus.[6] The harvesting process requires dissolving the blastocyst, bringing into question the moral and legal status of the human embryo.

Moral and Legal Status of the Human Embryo

Many religious perspectives consider the human fetus to constitute an individualized human entity. However, there is substantial debate regarding at which specific stage dignity is conferred in development (conception, primitive streak development, implantation, "quickening," or birth).[106–112] Recently, a less specific "developmental view" of moral status surfaced, meriting moral rights to the individual as consciousness and relationships develop.[113, 114]

Taking into account the many perspectives on the moral status of the human embryo and the scientific promises of a healthier tomorrow through stem cell technology, our society has attempted to define the legal status of the human embryo. In the United States, the first pillar was constructed in 1973 when the US Supreme Court ruled that a fetus is not a person in terms of constitutional protection (*Roe v Wade*, 410 US 113 [1973]). For a better examination of the decision's effect on research, the National Institutes of Health (NIH) imposed a moratorium on fetal research, and Congress founded the National Commission, charged to put together policy and guidelines on fetal research. Four months later, the commission published a report encouraging fetal research because of its potential, provided that the research risks to the fetus were minimal and were only those that would be accepted for a term fetus.[115] Thus, despite *Roe v Wade*, the commission extended protection to a fetus (just as to adult patients)

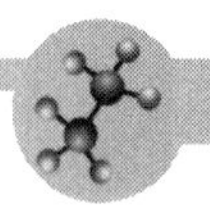

in research, including fetuses planned for elective abortion . . .

In the midst of this politically volatile federal sector support, private sector efforts to apply stem cell technology must navigate through a mosaicism of states rights laws. At the time of this writing, states have laws addressing research on embryos and fetuses, with 1, South Dakota, overtly banning hES research, punishable by a misdemeanor.[120] In contrast, California state law encourages research involving the use of human stem cells from any source and facilitates the voluntary donation of embryos for research.[121] Fetal research bans in Arizona, Illinois, Louisiana, and Utah were overturned by courts who found the laws vague. Furthermore, in states where no explicit laws have been passed to outlaw stem cell research, legal precedents regarding privacy and informed consent may threaten stem cell research[122] . . .

REFERENCES

6. Thomson JA, Itskovitz-Eldor J, Shapiro SS, et al. Embryonic stem cell lines derived from human blastocysts [published correction appears in *Science*. 1998;282:1827]. *Science*. 1998;282:1145-1147.
7. Henderson JK, Draper JS, Baillie HS, et al. Preimplantation human embryos and embryonic stem cells show comparable expression of stage-specific embryonic antigens. *Stem Cells*. 2002;20: 329-337.
8. Shamblott MJ, Axelman J, Wang S, et al. Derivation of pluripotent stem cells from cultured human primordial germ cells [published correction appears in *Proc Natl Acad Sci U S A*. 1999; 96;1162]. *Proc Natl Acad Sci U S A*. 1998;95:13726-13731.
9. Shamblott MJ, Axelman J, Littlefield JW, et al. Human embryonic germ cell derivatives express a broad range of developmentally distinct markers and proliferate extensively in vitro. *Proc Natl Acad Sci U S A*. 2001;98:113-118.

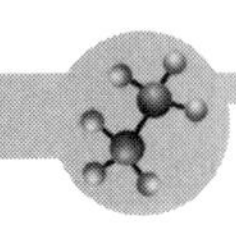

10. Amit M, Carpenter MK, Inokuma MS, et al. Clonally derived human embryonic stem cell lines maintain pluripotency and proliferative potential for prolonged periods of culture. *Dev Biol.* 2000;227:271-278.
11. Xu RH, Chen X, Li DS, et al. BMP4 initiates human embryonic stem cell differentiation to trophoblast. *Nat Biotechnol.* 2002;20:1261-1264.
12. Rohwedel J, Guan K, Hegert C, Wobus AM. Embryonic stem cells as an in vitro model for mutagenicity, cytotoxicity and embryotoxicity studies: present state and future prospects. *Toxicol In Vitro.* 2001;15:741-753.
13. Xu C, Police S, Rao N, Carpenter MK. Characterization and enrichment of cardiomyocytes derived from human embryonic stem cells. *Circ Res.* 2002;91:501-508.
14. Carpenter MK, Inokuma MS, Denham J, Mujtaba T, Chiu CP, Rao MS. Enrichment of neurons and neural precursors from human embryonic stem cells. *Exp Neurol.* 2001;172:383-397.
15. Assady S, Maor G, Amit M, Itskovitz-Eldor J, Skorecki KL, Tzukerman M. Insulin production by human embryonic stem cells. *Diabetes.* 2001;50:1691-1697.
16. Levenberg S, Golub JS, Amit M, Itskovitz-Eldor J, Langer R. Endothelial cells derived from human embryonic stem cells. *Proc Natl Acad Sci U S A.* 2002;99:4391-4396.
22. Campbell KH, McWhir J, Ritchie WA, Wilmut I. Sheep cloned by nuclear transfer from a cultured cell line. *Nature.* 1996;380:64-66.
23. Wilmut I, Schnieke AE, McWhir J, Kind AJ, Campbell KH. Viable offspring derived from fetal and adult mammalian cells [published correction appears in *Nature.* 1997;386:200]. *Nature.* 1997;385:810-813.
32. Hill JR, Burghardt RC, Jones K, et al. Evidence for placental abnormality as the major cause of mortality in first-trimester somatic cell cloned bovine fetuses. *Biol Reprod.* 2000;63:1787-1794.
33. De Sousa PA, King T, Harkness L, Young LE, Walker SK, Wilmut I. Evaluation of gestational deficiencies in cloned sheep fetuses and placentae. *Biol Reprod.* 2001;65:23-30.
34. Ogonuki N, Inoue K, Yamamoto Y, et al. Early death of mice cloned from somatic cells. *Nat Genet.* 2002;30:253-254.
35. Tamashiro KL, Wakayama T, Blanchard RJ, Blanchard DC, Yanagimachi R. Postnatal growth and behavioral development of mice cloned from adult cumulus cells. *Biol Reprod.* 2000;63:328-334.
36. Lanza RP, Cibelli JB, Blackwell C, et al. Extension of cell lifespan and telomere length in animals cloned from senescent somatic cells. *Science.* 2000;288:665-669.
37. Petersen BE, Bowen WC, Patrene KD, et al. Bone marrow as a potential source of hepatic oval cells. *Science.* 1999;284:1168-1170.

38. Grant MB, May WS, Caballero S, et al. Adult hematopoietic stem cells provide functional hemangioblast activity during retinal neovascularization. *Nat Med.* 2002;8:607-612.
56. Wagers AJ, Sherwood RI, Christensen JL, Weissman IL. Little evidence for developmental plasticity of adult hematopoietic stem cells. *Science.* 2002;297:2256-2259.
57. Castro RF, Jackson KA, Goodell MA, Robertson CS, Liu H, Shine HD. Failure of bone marrow cells to transdifferentiate into neural cells in vivo. *Science.* 2002;297:1299.
58. Habibian HK, Peters SO, Hsieh CC, et al. The fluctuating phenotype of the lymphohematopoietic stem cell with cell cycle transit. *J Exp Med.* 1998;188:393-398.
59. Krause DS, Theise ND, Collector MI, et al. Multi-organ, multilineage engraftment by a single bone marrow-derived stem cell. *Cell.* 2001;105:369-377.
60. Lemischka IR. Clonal, in vivo behavior of the totipotent hematopoietic stem cell. *Semin Immunol.* 1991;3:349-355.
61. Rajagopal J, Anderson WJ, Kume S, Martinez OI, Melton DA. Insulin staining of ES cell progeny from insulin uptake. *Science.* 2003;299:363.
62. Terada N, Hamazaki T, Oka M, et al. Bone marrow cells adopt the phenotype of other cells by spontaneous cell fusion. *Nature.* 2002;416:542-545.
63. Ying QL, Nichols J, Evans EP, Smith AG. Changing potency by spontaneous fusion. *Nature.* 2002;416:545-548.
106. Jeremiah 1:5 (KJ).
107. Athenagoras. Supplication for the Christians. Legatio 35. 177 AD.
108. Humanae Vitae: Encyclical Letter of His Holiness Pope Paul VI on the Regulation of Birth. Vatican City, Roman Catholic Church; July 25, 1968.
109. McCormick R, National Commission for the Protection of Human Subjects of Biomedical and Behavioral Research. Experimentation on the fetus: policy proposals. In: *Appendix to Report and Recommendations: Research on the Fetus.* Washington, DC: Government Printing Office; 1976. Publication DHEW (05)76-128.
110. Genesis 2:7 (KJ).
111. The Qur'an. Imam Muslim narrates from Ibn Mas'ud that he said that he heard the Messenger of Allah saying, "When 42 nights have passed on the nutfah (mixed male and female discharge of semen), Allah sends an angel to form it. He creates its hearing, vision, skin, flesh, and bones. Then the angel says, O Allah! Male or female?"
112. The Bhagavad Gita. Chapter 2, Sutra 20.
113. Steinbock B. *Life Before Birth: The Moral and Legal Status of Embryos and Fetuses.* New York, NY: Oxford University Press; 1992.

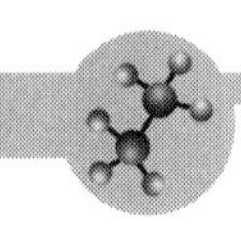

114. Warren MA. *Moral Status: Obligations to Persons and Other Living Things.* Oxford, England: Clarendon Press; 1997.
115. *Report and Recommendations: Research on the Fetus.* Washington, DC: National Commission for the Protection of Human Subjects of Biomedical and Behavioral Research, US Dept of Health, Education, and Welfare; 1975.
120. State of South Dakota, Medical Research, Chapter 14, Sections 34-14-16 to 34-14-20. Available at: http://nchla.org/cloning /southdakota.pdf. Accessibility verified July 10, 2003.
121. State of California, Health and Safety Code. §125115-125117 (September 22, 2002).
122. Andrews LB. State regulation of embryo stem cell research. In: National Bioethics Advisory Commission. *Ethical Issues in Human Stem Cell Research.* Volume II: Commissioned Papers. Rockville, Md: National Bioethics Advisory Commission; 2000: A-1-A-13.

Beyond tissue repair and regeneration, embryonic stem cells have uses for cloning. Cloning is the ability to create an entire organism from a single cell. It has been a topic long relegated to the ranks of science fiction (Star Trek, Woody Allen's Sleeper, *Ira Levin's* The Boys from Brazil*). However, in 1997, cloning moved from science fiction to science fact when Ian Wilmut and his colleagues successfully produced a clone of an adult sheep called Dolly. The cloning technique was called somatic cell nuclear transfer (SCNT). In SCNT, the nucleus of an adult cell is transferred to an oocyte, which was then stimulated to develop, thereby producing an organism that is genetically identical to the donor adult. Since*

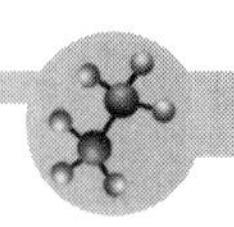

Dolly, many animals have been cloned for uses such as preserving endangered species or mass-producing animals (genetically engineered animals).

Despite successful cloning attempts and potential applications, cloning has many technical problems. Cloning techniques are inefficient, as few cloned embryos actually develop fully. Even when they develop, the cloned animals do not live normal, healthy lives. Keith Latham addresses these technical issues in the following article. —CCF

From "Cloning: Questions Answered and Unsolved"

by Keith E. Latham
***Differentiation*, 2004**

Cloning by adult somatic cell nuclear transfer has proven to be one of the most provocative achievements in recent scientific history. The successful production of the sheep Dolly, followed soon by the production of the cloned mouse Cumulina and then the birth of clones from other species using a variety of fetal and adult donor cell types, launched a frenzy of public and scientific debate about the cloning technology and how it should be applied. In the flood of lay and scientific commentary, it has at times been overlooked that the cloning technology is a new and emerging technology. Cloning is currently highly inefficient, but as with any

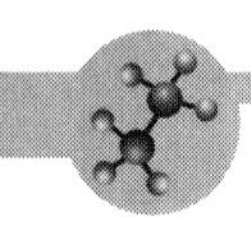

new technology it will likely improve to overcome this inefficiency, expand its applicability, and extend its scientific, commercial, and therapeutic value . . .

Timing of Nuclear Reprogramming and Its Effects on Cloned Embryos

Cloning by SCNT is a remarkable process in which a somatic cell nucleus is acted upon by the ooplasm via mechanisms that today remain entirely unknown. This effect of the ooplasm reflects the truly unique ability of the oocyte to create a functional embryonic genome from the two differentiated gamete (sperm and egg) genomes. In essence, cloning demands that the ooplasmic factors responsible for forming the embryonic genome act upon an alternate substrate. The degree of success in cloning is thus likely affected by the efficiency with which the ooplasmic factors operate on a foreign genome.

It has been widely assumed that the nuclear reprogramming that makes cloning possible occurs during the period immediately following nuclear transfer. In other words, it is assumed that the somatic cell nuclei are mysteriously converted to embryonic genomes, which then passively progress down the developmental path leading to a new individual. Numerous studies in fertilized embryos, however, have made it clear that the embryonic genome is first formed in the zygote, but is then drastically altered to set it on its path of totipotentiality.

Recent results from cloning studies indicate that the same situation likely applies during cloning, with the oocyte initiating the process of reprogramming, but

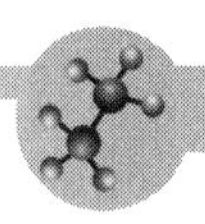

with the reprogramming process continuing as development progresses. In other words, reprogramming does not occur within the hours immediately following SCNT, but occurs progressively during cleavage, and likely continues after implantation . . .

Nuclear-Cytoplasmic Compatibility in Cloned Embryos Within and Between Species

Because early embryogenesis requires correct, coordinated interactions between the ooplasm and the nucleus, and because sustained metabolism relies upon effective cooperation between the nucleus and mitochondria, the question arises as to whether SCNT can provide insight into the overall degree of compatibility between nuclei, ooplasm, and mitochondria of different species, subspecies, or strains of animals.

Successful inter-species nuclear transfer has been performed between closely related species in several genera. Cloning was used for the genetic rescue of an endangered species of sheep, *Ovis orientalis musimon*, by SCNT into oocytes from domestic sheep, and then transferring those cloned embryos into the uteri of domestic sheep as foster mothers. Somatic cell nuclei from fetal buffalo (*Bubalus bubalis*) supported the production of blastocyst stage embryos when transplanted to bovine oocytes as efficiently as when transplanted to buffalo oocytes. Similarly, viable progeny have been produced after transplanting nuclei from *Bos indicus* cells to *Bos taurus* oocytes.

More extreme cross-genera combinations have not produced viable progeny to date, but have resulted in

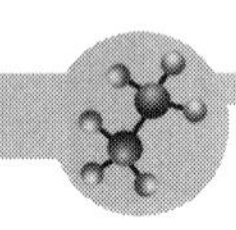

successful early development. Rhesus monkey fibroblast nuclei supported blastocyst formation after transfer to rabbit oocytes. The transfer of cat fibroblast nuclei to rabbit oocytes produced embryos that undergo the first few cleavage divisions before arresting. The rate of cleavage of the cat-rabbit NT embryos was slower than that of control rabbit-rabbit NT embryos, and the cat-rabbit embryos displayed an altered culture medium preference, again pointing to an early, dominant effect of the somatic nucleus on embryo phenotype and culture medium preference. Mouse embryo fibroblast nuclei transferred to bovine oocytes failed to direct appropriate gene expression and development past the eight-cell stage, indicating an inability of the bovine ooplasm to direct efficient activity of the mouse nucleus. Finally, giant panda nuclei have been introduced into rabbit oocytes and transferred to the reproductive tracts of either rabbits or cats. Embryo transfer to rabbit foster mothers did not result in pregnancy, but embryo transfer to feline foster mothers resulted in early stage fetuses, which were confirmed genetically to be from panda-rabbit NT embryos. It is striking that fetal development was achieved by inter-species hybrid NT embryos, and, furthermore, in the uterus of a third species . . .

Resetting Cellular Replication Potential

Implicit in the production of a new life by cloning with adult cell nuclei is the premise that clonal development might lead to the establishment of a greater replicative potential than the donor cell would have possessed

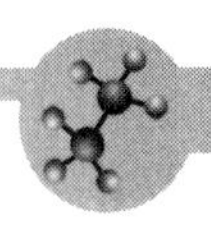

originally. Whether this indeed is the case depends upon a variety of factors, and the answer seems likely to be complicated. One issue is the nature of the donor cell from which the nucleus is obtained. Stem cells may possess the capacity to activate telomerase expression and restore telomere length to counter aging effects evident in differentiated cells. This may vary with cell type (e.g., hematopoietic stem cell versus muscle satellite cell.) Thus, whether a rare, surviving clone is derived from a differentiated cell of limited replicative potential, or from a contaminating stem cell that was able to manifest greater telomerase expression and replicative potential, could dramatically affect cloned animal lifespan and apparent telomere length. Additionally, because replicative potential varies with species and adult size, the replicative potential of cells in cloned embryos could vary along these parameters . . .

Problems to Address

Some important questions remaining are: why is cloning so inefficient? How can the efficiency of cloning be improved? Addressing these questions is likely to provide significant new insight into normal embryogenesis, basic cell biology, and our basic knowledge of gene regulatory mechanisms. In the process, advances in these areas may lead to significant improvements in the cloning technology.

One issue that has not been addressed in depth is the role of ooplasmic factors in normal embryos, and what is the effect on embryogenesis of removing oocyte components when constructing clones. It has become

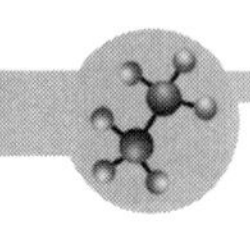

clear during the past 2 years that many aberrant characteristics of cloned embryos can be abrogated by leaving the oocyte spindle-chromosome complex (SCC) in the oocyte. A striking illustration of the regulative properties of the SCC relates to the capacity of the ooplasm to direct the exchange of H1 linker histones after SCNT. We recently found that the ooplasm normally directs the rapid removal of somatic H1 and its replacement with the oocyte-specific form. The capacity of the oocyte to direct this exchange is lost soon after oocyte activation. The SCC regulates the timing of this loss. This indicates that the SCC likely serves functions other than simply directing chromosome segregation, and may be involved in protein transport and localization, cell cycle control, and the control of gene transcription. Little is known about the composition and regulation of the meiotic spindle, and in particular the meiotic spindle of oocytes, which undergo a prolonged arrest in metaphase. Further knowledge in this area could reveal the molecular consequences of SCC removal, which thus far has been an obligate first step in the cloning procedure.

Another question to address is why imprinting information is disrupted in cloned embryos. Aberrant patterns of DNA methylation and imprinted gene expression have been reported for cloned embryos and progeny in different species, and in some reports it appears that the extra-embryonic tissues may be particularly susceptible to such defects. The basis for this difference between extra-embryonic and embryonic tissues warrants closer scrutiny. We have observed

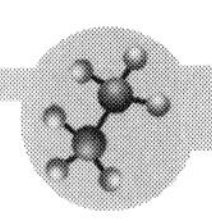

disruptions of DNA methylation without immediate effects on allele-specific expression in cloned blastocysts, revealing that our understanding of how DNA methylation represses imprinted allele expression is still incomplete. Further studies in cloned embryos will provide the opportunity to explore imprinted control of gene expression, and the loss/retention of imprinting information in various tissues that differ with respect to mono-allelic or bi-allelic expression.

Last, and most important, the very fundamental question remains: what is nuclear reprogramming? This term is widely used and generally accepted to indicate the termination of one gene expression program (e.g., donor cell) and the initiation of another program (e.g., embryonic). The molecular basis of reprogramming during cloning is entirely unknown. Although numerous correlates have been examined, such as effects on nuclear matrix proteins, nuclear envelope components, nucleolar components, and histone proteins, the actual factors that are responsible for guiding the specific repression and activation of genes in the cloned embryo have not been identified. Additionally, it is clear that reprogramming is only initiated in the oocyte, and then continues during cleavage and subsequent development, raising the question of what reprogramming factors exist in the oocyte, and what specific changes must occur in the oocyte to set the stage for continued reprogramming later. During normal embryogenesis, DNA replication appears to play a critical role in creating an embryonic genome that can be regulated correctly. This raises the questions: How

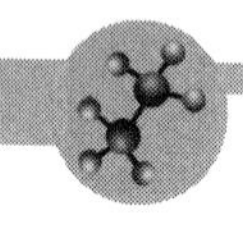

does DNA replication contribute to reprogramming in cloned embryos? Do the first one or two rounds of replication accomplish a lesser degree of reprogramming in clones as compared with normal fertilized embryos, due to differences in origin and initial chromatin state of the genomes? Given that reprogramming is initially incomplete and/or slow, the questions arise: what determines gene accessibility to reprogramming? Which genes become silenced or activated at which times? How does this relate to gene regulation during normal development? These are all basic questions that can be uniquely addressed using the SCNT technology, making this a valuable and exciting approach for addressing fundamental questions in biology.

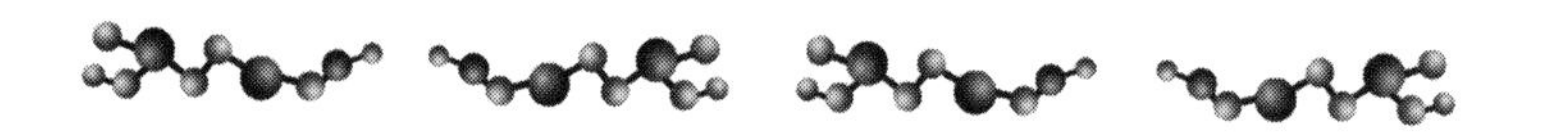

Even modern biotechnology uses what nature has already developed, albeit perhaps in different ways than intended. For example, as discussed previously, an undifferentiated stem cell can be enticed to develop into a liver cell or brain cell. Such a stem cell can also be transplanted with a nucleus of an adult somatic cell and cloned into an organism. Also, a gene for a jellyfish protein can be placed in a mouse embryo so that the adult mouse glows in the dark. In all of these cases, scientists have used natural starting materials (stem cells, genes, and proteins) that already

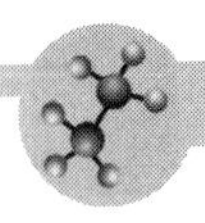

exist. However, a true example of bioengineering would be to create a biological organism that is entirely man-made. Imagine specifying all of the qualities that you might wish in an organism to do some specialized task, giving those specifications to a bioengineer, and having him or her manufacture your specified organism. In the final article, science writer W. Wayt Gibbs discusses how scientists are assembling genes and DNA "circuits" to make truly novel, synthetic life-forms and founding a new area of biology called synthetic biology. —*CCF*

"Synthetic Life"
by W. Wayt Gibbs
***Scientific American*, May 2004**

Biologists are crafting libraries of interchangeable DNA parts and assembling them inside microbes to create programmable, living machines.

Evolution is a wellspring of creativity; 3.6 billion years of mutation and competition have endowed living things with an impressive range of useful skills. But there is still plenty of room for improvement. Certain microbes can digest the explosive and carcinogenic chemical TNT, for example—but wouldn't it be handy if they glowed as they did so, highlighting the location of buried land mines or contaminated soil? Wormwood shrubs generate a potent medicine against malaria but only in trace quantities that are expensive to extract.

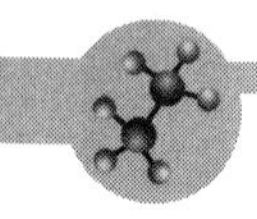

How many millions of lives could be saved if the compound, artemisinin, could instead be synthesized cheaply by vats of bacteria? And although many cancer researchers would trade their eyeteeth for a cell with a built-in, easy-to-read counter that ticks over reliably each time it divides, nature apparently has not deemed such a thing fit enough to survive in the wild.

It may seem a simple matter of genetic engineering to rewire cells to glow in the presence of a particular toxin, to manufacture an intricate drug, or to keep track of the cells' age. But creating such biological devices is far from easy. Biologists have been transplanting genes from one species to another for 30 years, yet genetic engineering is still more of a craft than a mature engineering discipline.

"Say I want to modify a plant so that it changes color in the presence of TNT," posits Drew Endy, a biologist at the Massachusetts Institute of Technology. "I can start tweaking genetic pathways in the plant to do that, and if I am lucky, then after a year or two I may get a 'device'—one system. But doing that once doesn't help me build a cell that swims around and eats plaque from artery walls. It doesn't help me grow a little microlens. Basically the current practice produces pieces of art."

Endy is one of a small but rapidly growing number of scientists who have set out in recent years to buttress the foundation of genetic engineering with what they call synthetic biology. They are designing and building living systems that behave in predictable ways, that use interchangeable parts, and in some cases that operate with an expanded genetic code, which allows them to do things that no natural organism can.

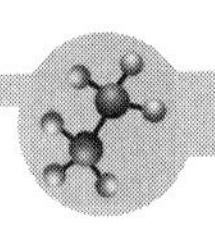

This nascent field has three major goals: One, learn about life by building it, rather than by tearing it apart. Two, make genetic engineering worthy of its name—a discipline that continuously improves by standardizing its previous creations and recombining them to make new and more sophisticated systems. And three, stretch the boundaries of life and of machines until the two overlap to yield truly programmable organisms. Already TNT-detecting and artemisinin-producing microbes seem within reach. The current prototypes are relatively primitive, but the vision is undeniably grand: think of it as Life, version 2.0.

A Light Blinks On

The roots of synthetic biology extend back 15 years to pioneering work by Steven A. Benner and Peter G. Schultz. In 1989 Benner led a team at ETH Zurich that created DNA containing two artificial genetic "letters" in addition to the four that appear in life as we know it. He and others have since invented several varieties of artificially enhanced DNA. So far no one has made genes from altered DNA that are functional—transcribed to RNA and then translated to protein form—within living cells. Just within the past year, however, Schultz's group at the Scripps Research Institute developed cells (containing normal DNA) that generate unnatural amino acids and string them together to make novel proteins.

Benner and other "old school" synthetic biologists see artificial genetics as a way to explore basic questions, such as how life got started on earth and what

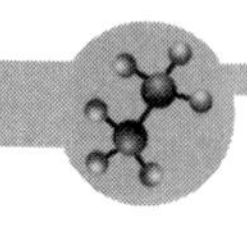

forms it may take elsewhere in the universe. Interesting as that is, the recent buzz growing around synthetic biology arises from its technological promise as a way to design and build machines that work inside cells. Two such devices, reported simultaneously in 2000, inspired much of the work that has happened since.

Both devices were constructed by inserting selected DNA sequences into *Escherichia coli*, a normally innocuous bacterium in the human gut. The two performed very different functions, however. Michael Elowitz and Stanislaus Leibler, then at Princeton University, assembled three interacting genes in a way that made the E. coli blink predictably, like microscopic Christmas tree lights. Meanwhile James J. Collins, Charles R. Cantor and Timothy S. Gardner of Boston University made a genetic toggle switch. A negative feedback loop—two genes that interfere with each other—allows the toggle circuit to flip between two stable states. It effectively endows each modified bacterium with a rudimentary digital memory.

To engineering-minded biologists, these experiments were energizing but also frustrating. It had taken nearly a year to create the toggle switch and about twice that time to build the flashing microbes. And no one could see a way to connect the two devices to make, for example, blinking bacteria that could be switched on and off.

"We would like to be able to routinely assemble systems from pieces that are well described and well behaved," Endy remarks. "That way, if in the future someone asks me to make an organism that, say, counts

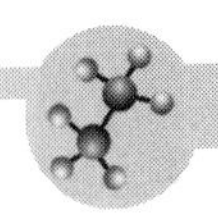

to 3,000 and then turns left, I can grab the parts I need off the shelf, hook them together and predict how they will perform." Four years ago parts such as these were just a dream. Today they fill a box on Endy's desk.

Building with BioBricks

"These are genetic parts," Endy says as he holds out a container filled with more than 50 vials of clear, syrupy fluid. "Each of these vials contains copies of a distinct section of DNA that either performs some function on its own or can be used by a cell to make a protein that does something useful. What is important here is that each genetic part has been carefully designed to interact well with other parts, on two levels." At a mechanical level, individual BioBricks (as the M.I.T. group calls the parts) can be fabricated and stored separately, then later stitched together to form larger bits of DNA. And on a functional level, each part sends and receives standard biochemical signals. So a scientist can change the behavior of an assembly just by substituting a different part at a given spot.

"Interchangeable components are something we take for granted in other kinds of engineering," Endy notes, but genetic engineering is only beginning to draw on the power of the concept. One advantage it offers is abstraction. Just as electrical engineers need not know what is inside a capacitor before they use it in a circuit, biological engineers would like to be able to use a genetic toggle switch while remaining blissfully ignorant of the binding coefficients and biochemical makeup of the promoters, repressors, activators, inducers and other genetic elements that make the switch work. One of the vials in Endy's box, for example, contains an inverter BioBrick (also called a NOT operator). When its input signal is high, its output signal is low, and vice versa. Another BioBrick performs a Boolean AND function, emitting an output signal only when it receives high levels of both its inputs. Because the two parts work with compatible signals, connecting them creates a NAND (NOT AND) operator. Virtually any binary computation can be performed with enough NAND operators.

Beyond abstraction, standardized parts offer another powerful advantage: the ability to design a functional genetic system without knowing exactly how to make it. Early last year a class of 16 students was able in one month to specify four genetic programs to make groups of *E. coli* cells flash in unison, as fireflies sometimes do. The students did not know how to create DNA sequences, but they had no need to. Endy hired a DNA-synthesis company to manufacture the 58 parts called for in their designs. These new BioBricks were then added to M.I.T.'s Registry of Standard Biological Parts. That online database today lists more than 140 parts, with the number growing by the month.

Hijacking Cells

As useful as it has been to apply the lessons of other fields of engineering to genetics, beyond a certain point the analogy breaks down. Electrical and mechanical machines are generally self-contained. That is true for a select few genetic devices: earlier this year, for example,

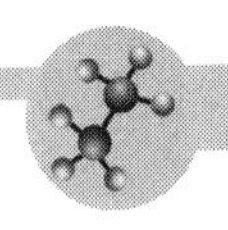

Milan Stojanovic of Columbia University contrived test tubes of DNA-like biomolecules that play a chemical version of tic-tac-toe. But synthetic biologists are mainly interested in building genetic devices within living cells, so that the systems can move, reproduce and interact with the real world. From a cell's point of view, the synthetic device inside it is a parasite. The cell provides it with energy, raw materials and the biochemical infrastructure that decodes DNA to messenger RNA and then to protein.

The host cell, however, also adds a great deal of complexity. Biologists have invested years of work in computer models of *E. coli* and other single-celled organisms [see "Cybernetic Cells," SCIENTIFIC AMERICAN, August 2001]. And yet, acknowledges Ron Weiss of Princeton, "if you give me the DNA sequence of your genetic system, I can't tell you what the bacteria will do with it." Indeed, Endy recalls, "about half of the 60 parts we designed in 2003 initially couldn't be synthesized because they killed the cells that were copying them. We had to figure out a way to lower the burden that carrying and replicating the engineered DNA imposed on the cells." (Eventually 58 of the 60 parts were produced successfully.)

One way to deal with the complexity added by the cells' native genome is to dodge it: the genetic device can be sequestered on its own loop of DNA, separate from the chromosome of the organism. Physical separation is only half the solution, however, because there are no wires in cells. Life runs on "wetware," with many protein signals simply floating randomly from

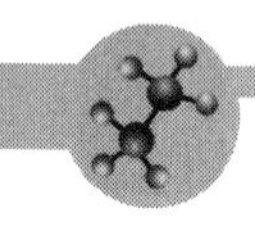

one part to another. "So if I have one inverter over here made out of proteins and DNA," Endy explains, "a protein signal meant for that part will also act on any other instance of that inverter anywhere else in the cell," whether it lies on the artificial loop or on the natural chromosome.

One way to prevent crossed signals is to avoid using the same part twice. Weiss has taken this approach in constructing a "Goldilocks" genetic circuit, one that lights up when a target chemical is present but only when the concentration is not too high and not too low. Tucked inside its various parts are four inverters, each of which responds to a different protein signal. But this strategy makes it much more difficult to design parts that are truly interchangeable and can be rearranged.

Endy is testing a solution that may be better for some systems. "Our inverter uses the same components [as one of Weiss's], just arranged differently," Endy says. "Now the input is not a protein but rather a rate, specifically the rate at which a gene is transcribed. The inverter responds to how many messenger RNAs are produced per second. It makes a protein, and that protein determines the rate of transcription going out [by switching on a second gene]. So I send in TIPS—transcription events per second—and as output, I get TIPS. That is the common currency, like a current in an electrical circuit." In principle, the inverter could be removed and replaced with any other BioBrick that processes TIPS. And TIPS signals are location-specific, so the same part can be used at several places in a circuit without interference.

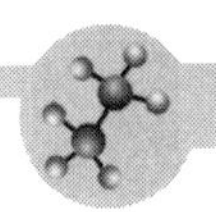

The TIPS technique will be tested by a new set of genetic systems designed by students who took a winter course at M.I.T. this past January. The aim this year was to reprogram cells to work cooperatively to form patterns, such as polka dots, in a petri dish. To do this the cells must communicate with one another by secreting and sensing chemical nutrients.

"This year's systems were about twice the size of the 2003 projects," Endy says. It took 13 months to get the blinking *E. coli* designs built and into cells. But in the intervening year the inventory of BioBricks has grown, the speed of DNA synthesis has shot up, and the engineers have gained experience assembling genetic circuits. So Endy expects to have the 2004 designs ready for testing in just five months, in time to show off at the first synthetic biology conference, scheduled for this June.

Rewriting the Book of Life

The scientists who attend that conference will no doubt commiserate about the inherent difficulty of engineering a relatively puny stretch of DNA to work reliably within a cell that is constantly changing. Living machines reproduce, but as they do they mutate.

"Replication is far from perfect. We've built circuits and seen them mutate in half the cells within five hours," Weiss reports. "The larger the circuit is, the faster it tends to mutate." Weiss and Frances H. Arnold of the California Institute of Technology have evolved circuits with improved performance using multiple rounds of mutation followed by selection of those cells

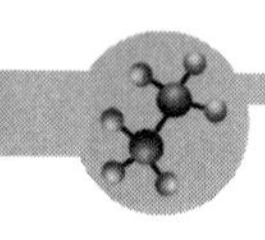

most fit for the desired task. But left unsupervised, evolution will tend to break genetic machines.

"I would like to make a genetically encoded device that accepts an input signal and simply counts: 1, 2, 3, . . . up to 256," Endy suggests. "That's not much more complex than what we're building now, and it would allow you to quickly and precisely detect certain types of cells that had lost control of their reproduction and gone cancerous. But how do I design a counter so that the design persists when the machine makes copies of itself that contain mistakes? I don't have a clue. Maybe we have to build in redundancy—or maybe we need to make the function of the counter somehow good for the cell."

Or perhaps the engineers will have to understand better how simple forms of life, such as viruses, have solved the problem of persistence. Synthetic biology may help here, too. Last November, Hamilton O. Smith and J. Craig Venter announced that their group at the Institute for Biological Energy Alternatives had re-created a bacteriophage (a virus that infects bacteria) called phiX174 from scratch, in just two weeks. The synthetic virus, Venter said, has the same 5,386 base pairs of DNA as the natural form and is just as active.

"Synthesis of a large chromosome is now clearly in reach," said Venter, who for several years led a project to identify the minimal set of genes required for survival by the bacterium *Mycoplasma genitalium*. "What we don't know is whether we can insert that chromosome into a cell and transform the cell's operating system to work off the new chromosome. We will have

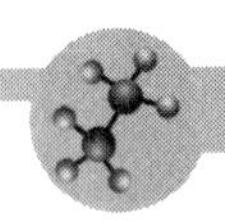

to understand life at its most basic level, and we're a long way from doing that."

Re-creating a virus letter-for-letter does not reveal much about it, but what if the genome were dissected into its constituent genes and then methodically put back together in a way that makes sense to human engineers? That is what Endy and colleagues are doing with the T7 bacteriophage. "We've rebuilt T7—not just resynthesized it but reengineered the genome and synthesized that," Endy reports. The scientists are separating genes that overlap, editing out redundancies, and so on. The group has completed about 11.5 kilobases so far and expects to finish the remaining 30,000 base pairs by the end of 2004.

Beta-Testing Life 2.0

Synthetic biologists have so far built living genetic systems as experiments and demonstrations. But a number of research laboratories are already working on applications. Martin Fussenegger and his colleagues at ETH Zurich have graduated from bacteria to mammals. Last year they infused hamster cells with networks of genes that have a kind of volume control: adding small amounts of various antibiotics turned the output of the synthetic genes to low, medium or high. Controlling gene expression in this way could prove quite handy for gene therapies and the manufacture of pharmaceutical proteins.

Living machines will probably find their first uses for jobs that require sophisticated chemistry, such as detecting toxins or synthesizing drugs. Last year

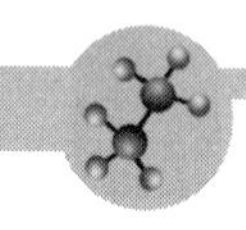

Homme W. Hellinga of Duke University invented a way to redesign natural sensor proteins in *E. coli* so that they would latch onto TNT or any other compound of interest instead of their normal targets. Weiss says that he and Hellinga have discussed combining his Goldilocks circuit with Hellinga's sensor to make land-mine detectors.

Jay Keasling, who recently founded a synthetic biology department at Lawrence Berkeley National Laboratory (LBNL), reports that his group has engineered a large network of wormwood and yeast genes into *E. coli*. The circuit enables the bacterium to fabricate a chemical precursor to artemisinin, a next-generation antimalarial drug that is currently too expensive for the parts of the developing world that need it most.

Keasling says that three years of work have increased yields by a factor of one million. By boosting the yields another 25- to 50-fold, he adds, "we will be able to produce artemisinin-based dual cocktail drugs to the Third World for about one tenth the current price." With relatively simple modifications, the bioengineered bacteria could be altered to produce expensive chemicals used in perfumes, flavorings and the cancer drug Taxol.

Other scientists at LBNL are using *E. coli* to help dispose of nuclear waste as well as biological and chemical weapons. One team is modifying the bacteria's sense of "smell" so that the bugs will swim toward a nerve agent, such as VX, and digest it. "We have engineered *E. coli* and *Pseudomonas aeruginosa* to precipitate heavy metals, uranium and plutonium on their

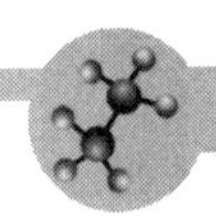

cell wall," Keasling says. "Once the cells have accumulated the metals, they settle out of solution, leaving cleaned wastewater."

Worthy goals, all. But if you become a touch uneasy at the thought of undergraduates creating new kinds of germs, of private labs synthesizing viruses, and of scientists publishing papers on how to use bacteria to collect plutonium, you are not alone.

In 1975 leading biologists called for a moratorium on the use of recombinant-DNA technology and held a conference at the Asilomar Conference Grounds in California to discuss how to regulate its use. Self-policing seemed to work: there has yet to be a major accident with genetically engineered organisms. "But recently three things have changed the landscape," Endy points out. "First, anyone can now download the DNA sequence for anthrax toxin genes or for any number of bad things. Second, anyone can order synthetic DNA from offshore companies. And third, we are now more worried about intentional misapplication."

So how does society counter the risks of a new technology without also denying itself all the benefits? "The Internet stays up because there are more people who want to keep it running than there are people who want to bring it down," Endy suggests. He pulls out a photograph of the class he taught last year. "Look. The people in this class are happy and building nice, constructive things, as opposed to new species of virus or new kinds of bioweapons. Ultimately we deal with the risks of biological technology by creating a society that can use the technology constructively."

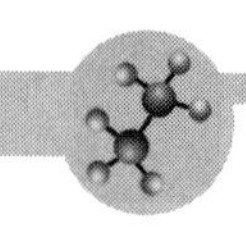

But he also believes that a meeting to address potential problems makes sense. "I think," he says, "it would be entirely appropriate to convene a meeting like Asilomar to discuss the current state and future of biological technology." This June, as leaders in the field meet to share their latest ideas about what can now be created, perhaps they will also devote some thought to what shouldn't.

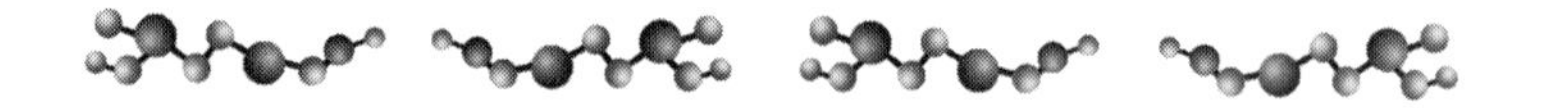

Conclusion

In this anthology, we have provided articles that address the following major questions or issues currently arising in developmental biology:

- *Cells and tissues can become specialized for different functions such as contraction. Although it is generally believed that specialized cells can no longer reproduce, these tissues can be made to regenerate. Furthermore, there are exceptions (such as the marine alga Caulerpa) to the widely accepted concept that all large organisms had to be multicelled with specialized tissues.*
- *Development involves differential gene regulation (e.g., RNAi), biochemical signals (morphogens), and various cellular processes*

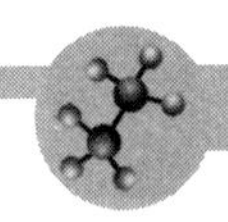

(cell division, changes in cell shape, cell migration, cell differentiation, and programmed cell death).

- *Various tissues (e.g., liver) can regenerate, which involves epithelial and/or stem cell recruitment and differentiation. The concepts learned about cell and tissue differentiation may provide new insights into cancer.*
- *Human ingenuity can apply the lessons learned from studying cell and tissue differentiation to regenerate or manufacture diseased and failing organs, to create whole organisms, and to create new synthetic life-forms.*

Throughout this anthology, we have used text excerpts of the original articles to provide information about the key concepts of growth and development of specialized cells, tissues, and organs. We refer you to the original texts for bibliographies, figures, and further information. We hope that this anthology has stimulated your interest in this fascinating field of biology. —*CCF*

Web Sites

Due to the changing nature of Internet links, the Rosen Publishing Group, Inc., has developed an on-line list of Web sites related to the subject of this book. This site is updated regularly. Please use this link to access the list:

http://www.rosenlinks.com/cdfb/scto

For Further Reading

Freeman, M., and J. B. Gurdon. "Regulatory Principles of Developmental Signaling." *Annual Review of Cell and Developmental Biology*, Vol. 18, 2002, pp. 515–539.

Gilbert, Scott. F. *Developmental Biology*, 7th ed. Sunderland, MA: Sinauer Associates, 2003.

Hombria, James C. G., and Bridget Lovegrove. "Beyond Homeosis—HOX Function in Morphogenesis and Organogenesis." *Differentiation*, Vol. 71, 2003, pp. 461–476.

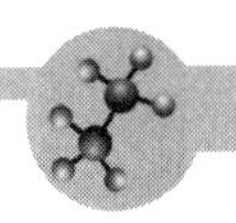

Horiguchi, Gorou. "RNA Silencing in Plants." *Differentiation*, Vol. 72, 2004, pp. 65–73.

Irvine, Kenneth D., and Cordelia Rauskolb. "Boundaries in Development: Formation and Function." *Annual Review Cell Developmental Biology*, Vol. 17, 2001, pp. 189–214.

Lam, Eric. "Controlled Cell Death, Plant Survival and Development." *Nature Reviews Molecular Cell Biology*, Vol. 5, 2004, pp. 305–315.

Lanza, Robert, and Nadia Rosenthal. "The Stem Cell Challenge." *Scientific American*, June 2004, pp. 93–99.

Nature Milestones. *Milestones in Development*. London, UK: Nature Publishing Group, 2004. Available at Web site (http://www.nature.com/milestones/development).

Renehan, Andrew G., Catherine Booth, and Christopher S. Potten. "What Is Apoptosis, and Why Is It Important?" *British Medical Journal* (BMJ),Vol. 322, June 23, 2001, pp. 1536–1538.

Tabata, Tetsuya, and Yuki Takei. "Morphogens, Their Identification and Regulation." *Development*,Vol. 131, 2004, pp. 703–712.

Tosh, David, and Jonathan M. W. Slack. "How Cells Change Their Phenotype." *Nature Reviews Molecular Cell Biology*, Vol. 3, 2002, pp. 187–194.

Bibliography

Belmonte, Juan C. Izpisúa. "How the Body Tells Left from Right." *Scientific American*, June 1999, pp. 46–51.

Cogle, Christopher R., Steven M. Guthrie, Ronald C. Sanders, et al. "An Overview of Stem Cell Research and Regulatory Issues." *Mayo Clinic Proceedings*, Vol. 78, 2003, pp. 993–1003.

Gibbs, W. Wayt. "Synthetic Life." *Scientific American*, May 2004, pp. 75–81.

Hülskamp, Martin "Plant Trichomes: A Model for Cell Differentiation." *Nature Reviews Molecular Cell Biology*, Vol. 5, June 2004, pp. 471–480.

Ingber, Donald E. "Cancer as a Disease of Epithelial-Mesenchymal Interactions and Extracellular Matrix Regulation." *Differentiation*, Vol. 70, 2002, pp. 547–560.

Jacobs, William P. "*Caulerpa*." *Scientific American*, December 1994, pp. 100–105.

Latham, Keith E. "Cloning: Questions Answered and Unsolved." *Differentiation*, Vol. 72, 2004, pp. 11–22.

Lau, Nelson C., and David P. Bartel. "Censors of the Genome." *Scientific American*, August 2003, pp. 34–41.

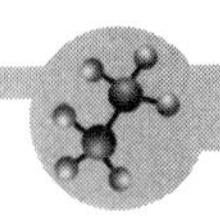

Lindsey, Keith, and Jennifer Topping. "On the Relationship Between the Plant Cell and the Plant." *Seminars in Cell & Developmental Biology*, Vol. 9, 1998, pp. 171–177.

Meier, Pascal, Andrew Finch, and Gerard Evan. "Apoptosis in Development." *Nature*, Vol. 407, October 2000, pp. 796–801.

Nüsslein-Volhard, Christiane. "Gradients That Organize Embryo Development." *Scientific American*, August 1996, pp. 54–61.

Oh, Seh-Hoon, Heather M. Hatch, and Bryon E. Petersen. "Hepatic Oval 'Stem' Cell in Liver Regeneration." *Seminars in Cell & Developmental Biology*, Vol. 13, 2002, pp. 405–409.

Potter, John D. "Morphostats: A Missing Concept in Cancer Biology." *Cancer Epidemiology, Biomarkers & Prevention*, Vol. 10, March 2001, pp. 161–170.

Riddle, Robert D., and Clifford J. Tabin. "How Limbs Develop." *Scientific American*, February 1999, pp. 74–79.

Sweeney, H. Lee. "Gene Doping." *Scientific American*, July 2004, pp. 63–69.

Williams, David. "Benefit and Risk in Tissue Engineering." *Materials Today*, May 2004, pp. 24–29.

Index

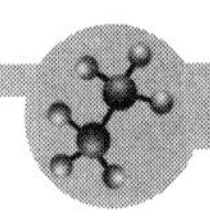

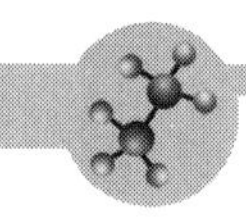

About the Editor

Craig C. Freudenrich earned a bachelor's degree in biology from West Virginia University and a doctorate in physiology from the University of Pittsburgh School of Medicine in Pennsylvania. He has twenty years of experience in biomedical research and eight years of science teaching experience in secondary schools. As a former senior editor (science, medicine, and the human body) for HowStuffWorks, he has written many articles on the human body and other science topics. He is currently a science writer and instructor for the Duke University Talent Identification Program in Durham, North Carolina.

Photo Credits

Front Cover: (Top, left inset) © Inmagine.com; (bottom left) © Pixtal/Superstock; (background) © Royalty Free/Corbis; (lower right inset) © The American Society for Biochemistry and Molecular Biology. Back Cover: (bottom inset) © Inmagine.com; (top) © Royalty Free/Corbis.

Designer: Geri Fletcher; Series Editor: Leigh Ann Cobb